Summer Counts!

2nd Edition

Grade 4→5

Thematic Reading, Language Arts, and Math Skills

Options™

ISBN 978-1-60161-929-7

OT104

Cover Image: Enrique Corts/Shannon Associates, LLC

Triumph Learning® 136 Madison Avenue, 7th Floor, New York, NY 10016

Printed in the United States of America.

10 9 8 7 6 5 4 3 2 1

Dear Parent,

Summer is a time for relaxing and having fun. It can also be a time for learning. *Summer Counts!* can help improve your child's understanding of important skills learned in the past school year while preparing him or her for the year ahead.

Summer Counts! provides grade-appropriate practice in subjects such as reading, language arts, vocabulary, and math. The ten theme-related chapters include activities and puzzles to motivate your child throughout the summer.

When working through the book, encourage your child to share his or her learning with you. You may want to tear out the answer key at the back of the book and use it to check your child's progress. With *Summer Counts!* your child will discover that learning happens anytime—even in the summer!

Apreciados padre,

El verano es una época para descansar y divertirse. También puede ser una época para aprender. *Summer Counts!* puede ayudar a que su hijo(a) mejore las destrezas importantes que aprendió el pasado año escolar al mismo tiempo que lo(a) prepara para el año que se aproxima.

Summer Counts! provee la práctica apropiada para cada grado en las asignaturas como la lectura, las artes del lenguaje y las matemáticas. Los diez capítulos temáticos incluyen actividades y rompecabezas que motivarán a su hijo(a) durante el verano.

Cuando trabaje con el libro, anime a su hijo(a) a que comparta lo que ha aprendido con Ud. Si Ud. desea puede desprender la página de las respuestas que aparece en la parte trasera del libro. Puede usar la misma para revisar el progreso de su hijo(a). ¡Con *Summer Counts!* su hijo(a) descubrirá que el aprendizaje puede ocurrir en cualquier momento—inclusive en el verano!

Table of Contents

CHAPTER 1

A Face Full of Fun

It started out like any other day for Shana. She got up and dressed in her favorite blue jeans and went down for breakfast.

It was the Fourth of July and Shana's mother had decorated the house in red, white, and blue ribbons. In the middle of the table stood the apple pie they had made yesterday for today's picnic. Shana could hear a flag flapping in the breeze outside. She could smell the chicken her mother was frying. She couldn't wait to get her hands on that chicken and pie.

Shana looked at the clock. It was only 8:00 A.M. The picnic didn't start until 1:00 P.M. What was she going to do until then? She was just finishing her cereal when her grandfather came into the kitchen.

"Grandpa," she said, "I love the Fourth of July. Will you tell me what you did to celebrate when you were my age?" She knew that her grandpa had grown up in the country and always had great stories to tell.

Grandpa stroked his beard and began. "The fun started with a parade through the center of town. The mayor, high school band, ladies' club, and even the prize pig were among those who marched through the streets of town. If you weren't in the parade, you made sure you were there to watch.

"Each family brought blankets and lots of great food. As the final entry marched past, everyone watching scooped up their belongings and followed the parade to an open field on the edge of town. It was in this field that the picnic took place. The kids played games and competed in contests. My two favorite contests were pie eating and sack racing. To win the pie-eating contest you had to eat the most pie in five minutes. This was not easy because you had to eat with your hands tied behind your back. It was face-first into the pie of your choice. In the sack race, you pulled a large cloth sack over your feet and legs and hopped to the finish line.

"The picnic always lasted late into the afternoon. By the time it was over, most of the children were in the creek swimming and the adults were stretched out on blankets under trees. Everyone rested up for the big dance later in the night. We did not have fireworks, so the night-time entertainment was a dance and party that everyone attended. Little kids came and grandparents came and everyone danced and played games until midnight."

When Grandpa had finished telling Shana about his memories of the Fourth of July, he looked over to see Shana face-first in the apple pie!

A Face Full of Fun

Directions Using what you have just read, answer the questions.

1. Think of parades you have seen in person or on television. How are parades today different from the one Grandpa described?

____

____

2. In the space below, draw a picture showing Shana and Grandpa at the end of the story. Make sure to include details from the story in your picture.

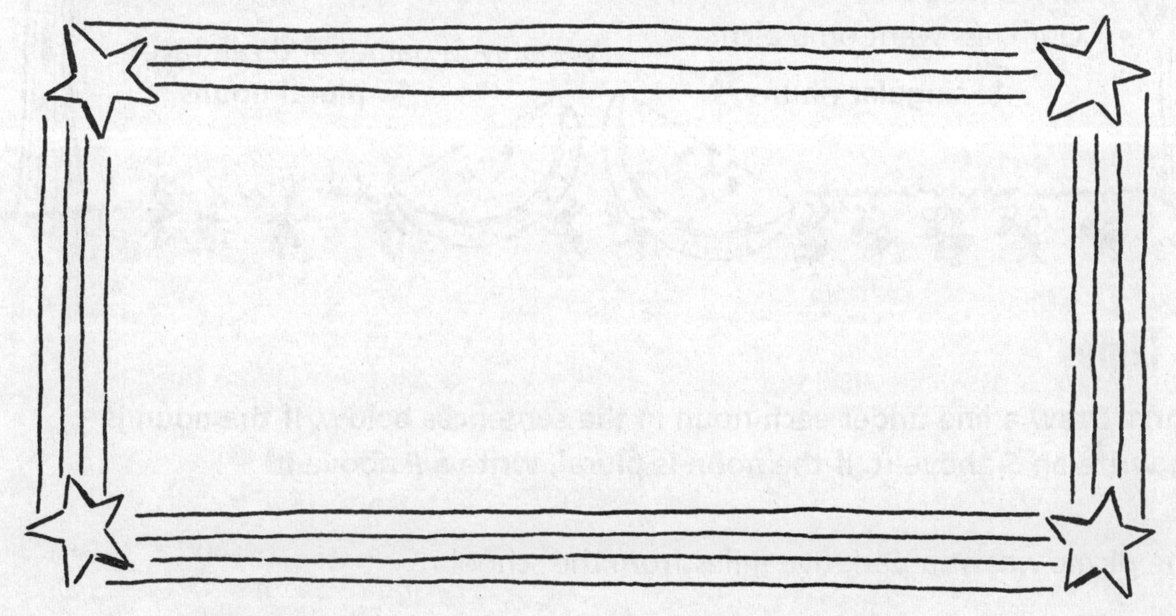

3. If you could have a picnic with anyone in the world from the past or present, who would it be and why?

____

____

____

4. Why did Shana's mother decorate the house in red, white, and blue ribbons?

____

____

____

Nouns: Singular and Plural

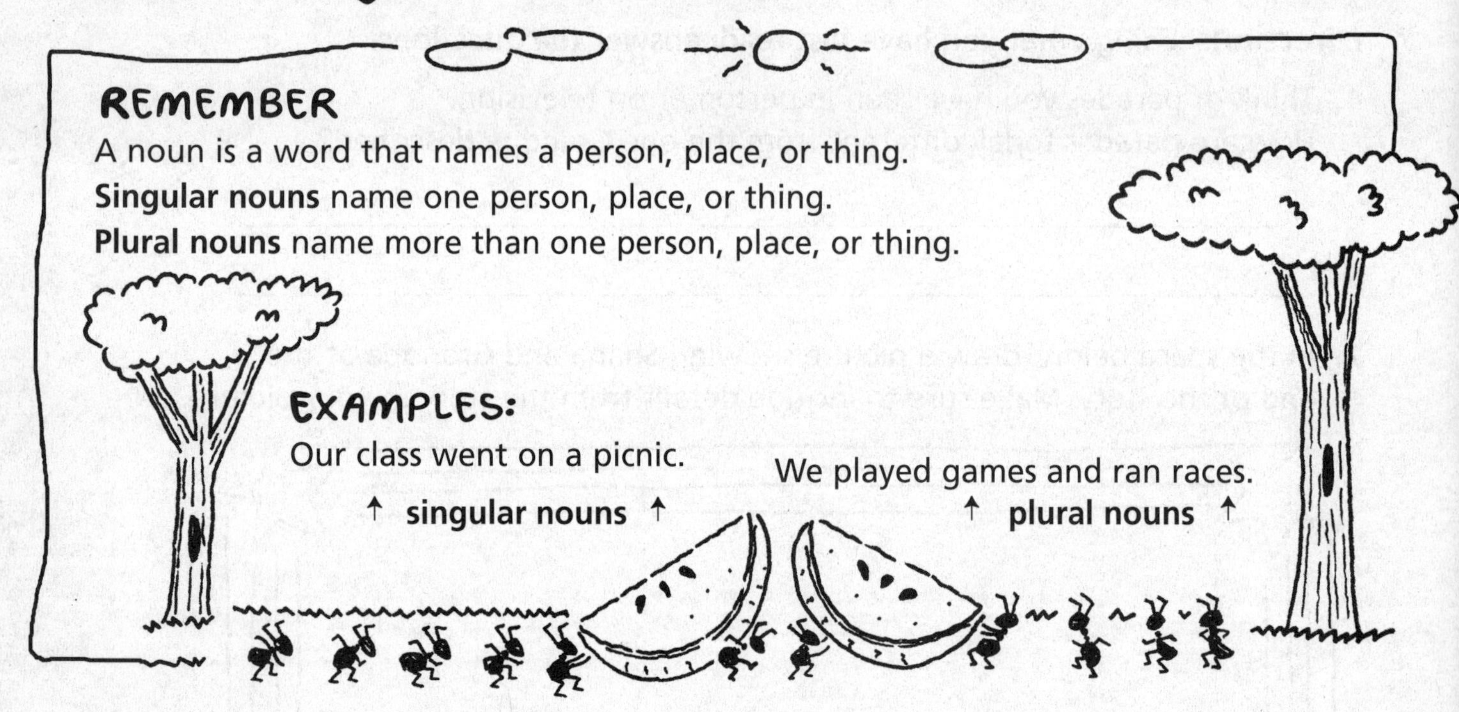

REMEMBER

A noun is a word that names a person, place, or thing.
Singular nouns name one person, place, or thing.
Plural nouns name more than one person, place, or thing.

EXAMPLES:

Our class went on a picnic.
↑ **singular nouns** ↑

We played games and ran races.
↑ **plural nouns** ↑

Picnic Time

Directions Draw a line under each noun in the sentences below. If the noun is singular, write an *S* above it. If the noun is plural, write a *P* above it.

1. The picnic was at a park five miles from the school.
2. We rode to the picnic on a bus.
3. The students brought sack lunches to eat.
4. Our teacher brought apples and a huge plastic jug of lemonade.
5. The park had a pond with geese and ducks in it.
6. Then the class took a walk on a trail.

Word Shapes

Directions Use words from the box below to fill in the box shapes.

chicken	charge	which
watch	rich	teacher

1. 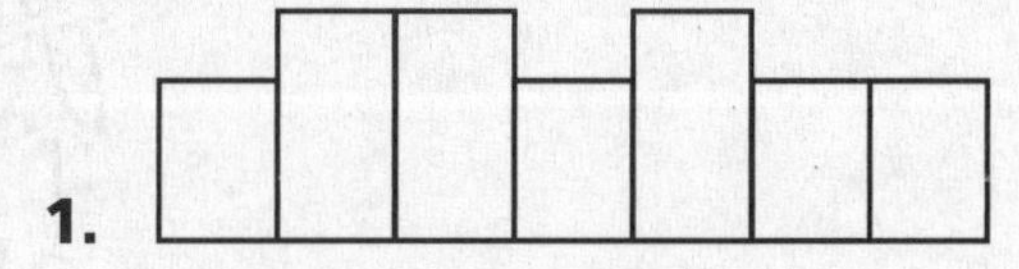

2. 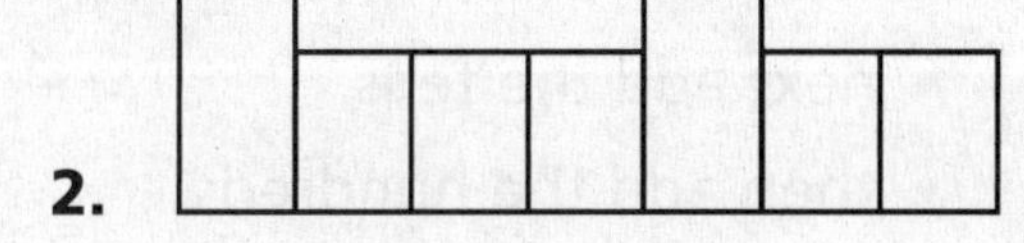

3. 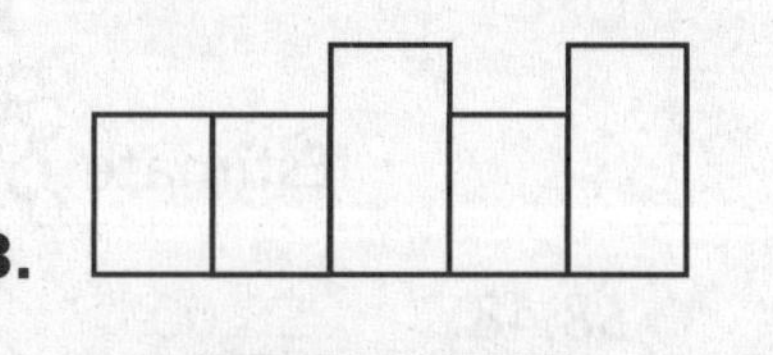

4. 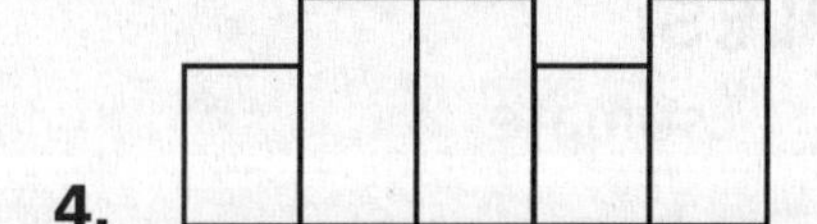

5. 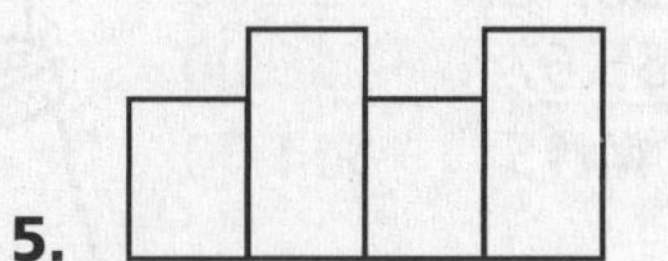

6. 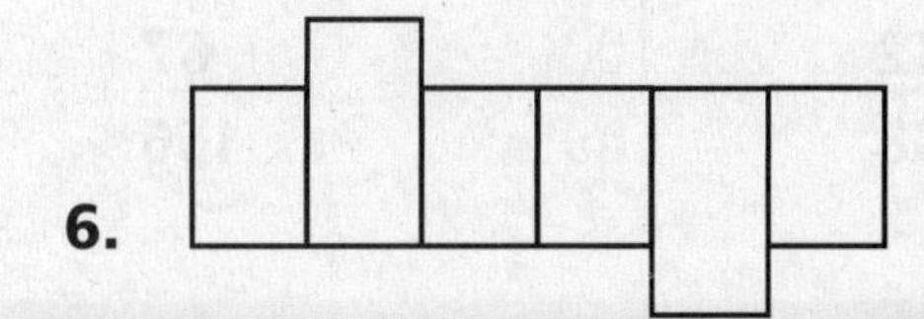

chicken	peach	reach	lunch	teacher	inches
children	such	watch			

The Class Outing

Directions Read the story below. Use the words from the box above to fill in the blanks.

It was **(7)** ____________________ a nice day today, our

(8) ____________________ decided that we could have a picnic

(9) ____________________. Our class went to the park with coolers and

picnic baskets full of food. I had cold fried **(10)** ____________________ and

(11) ____________________ pie. Other **(12)** ____________________

were playing on the swings. We didn't want to **(13)** ____________________.

We wanted to play too! So, after we ate, we had a race. I jumped as fast as I could to

(14) ____________________ the finish line first. I won by just a few short

(15) ____________________!

Adding Whole Numbers

When you add, you combine numbers or groups of objects. Adding can be thought of as a quick way of counting. Estimating before you add can also be helpful.

REMEMBER

To add numbers, start with the ones and work from right to left.

- First add the ones,
- next add the tens,
- then add the hundreds.

EXAMPLES:

	Estimate		Estimate		Estimate
426	400	89	90	$8.48	$8.00
+ 112	+ 100	+ 67	+ 70	+ $5.67	+ $6.00
538	**500**	**156**	**160**	**$14.15**	**$14.00**

It All Adds Up

Directions **Find the sum. Estimate as your first step. The first one is done for you.**

		Estimate			Estimate			Estimate
1.	43 + 24 67	40 + 20 60	**2.**	61 + 27		**3.**	82 + 43	
4.	184 + 41		**5.**	217 + 117		**6.**	624 + 741	
7.	$4.78 + $2.10		**8.**	$6.09 + $3.09		**9.**	$27.30 + $11.15	

Even and Odd

Directions Use whole numbers to solve.

1. The digits 0, 1, 2, 3, 4, 5, 6, 7, 8, and 9 are used to make all whole numbers. Of these, list the five even digits. (Hint: 0 is considered an even number.)

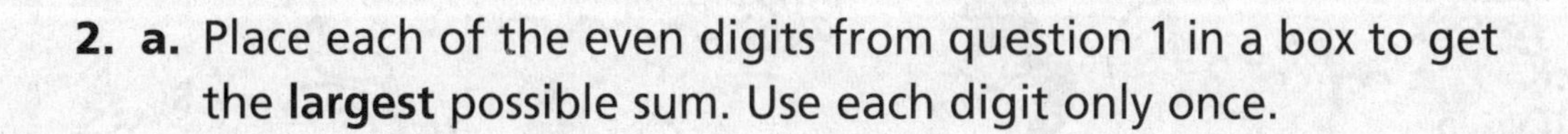

2. **a.** Place each of the even digits from question 1 in a box to get the **largest** possible sum. Use each digit only once.

   b. What is the largest possible sum? ____________

3. **a.** Place any digit (0 to 9) in the boxes below to get the **smallest** possible difference. Use each digit only once.

   b. What is the smallest possible difference? ____________

Picnic Word Search

Directions **Find and circle the words listed in the box below. The words can be hidden across, down, or diagonally.**

park	lunch	chicken	coolers	family	lemonade
class	laugh	ants	picnic	basket	playground

T	A	C	F	A	M	I	L	Y	P
P	I	T	O	L	B	Z	D	L	L
I	L	E	M	O	N	A	D	E	A
C	L	D	F	S	L	Q	P	F	Y
N	L	U	G	U	A	E	A	B	G
I	S	A	N	A	U	W	R	A	R
C	K	P	S	C	G	E	K	S	O
D	Z	P	O	S	H	U	P	K	U
H	C	M	C	H	I	C	K	E	N
S	W	A	N	T	S	Y	O	T	D

The Ants' Picnic

Yesterday was my birthday. Mom made a picnic in our backyard. Mom brought out my favorite foods. Then she went inside and left my younger brother, Kenny, to watch over the food while my friends and I played soccer.

Then it was time to eat, but when I looked at the food on the blanket there were ants all over the place! They were eating the food and crawling over the watermelon! "I'm sorry, Lana," said Kenny. "I didn't see them because I was watching your soccer game."

Mom came out and gathered up the food. She threw away most of it. Then Mom took everyone out for pizza. It wasn't a picnic, but it was still a great birthday party!

Directions Circle the letter of the best answer.

1. What was Kenny supposed to do?

A. bring the food outside

B. bring the food inside

C. prepare the food

D. watch over the food

2. Who is telling the story?

A. Mom **B.** Kenny

C. Lana **D.** Lana's friend

3. Which sentence does **not** describe Mom?

A. She is a calm person. **B.** She gets upset easily.

C. She solves problems. **D.** She is a caring person.

4. Which of the following is a plural noun?

A. birthday **B.** ants

C. backyard **D.** brother

CHAPTER 2

Venom and Victims

What kind of animal has eight legs (always), eight eyes (mostly), and makes many kinds of traps? The answer is one of the best predators around: the spider. Predators are animals that live by killing and eating other animals. The animals they eat are called prey.

Spiders eat mostly insects and other spiders. Some very large spiders can eat small birds! But before a spider eats its prey, it has to capture it. Spiders have more than one way to do that:

1. The spider hunts— that is, it runs after a victim and attacks it. Spiders that hunt are called wolf spiders. They hunt at night and live on the ground.
2. The spider hides, waiting for prey to come by. When it does, the spider runs out and attacks.
3. The spider makes a trap, waiting for prey to get caught. Most spiders make a trap.

There are many kinds of traps, such as a hole in the ground with a trap door over it. The most common type of trap is the web. Spiders weave their webs with silk from their bodies. The silk web is like a net in the air. When a fly or bee or other insect flies into the sticky net, it gets caught. Then the spider, who has been waiting in the web, runs over to the insect and bites it.

All spiders use venom to kill their prey. As they bite a victim with their fangs, venom travels through the fangs into the victim. The victim quickly dies.

Spiders cannot chew their food. They have to drink it. So, after killing their prey, they turn it into a watery soup. What would it be like to drink *that* kind of soup?

Venom and Victims

Directions Using what you have just read, answer the questions.

1. List three ways a spider captures food.

2. What is venom?

3. Why do you think a spider cannot chew its food?

4. The spider is prey for some animals. Name two animals that could be predators of the spider.

5. Is this selection real or make-believe? List three clues that help you know.

Possessive Nouns

REMEMBER

A **possessive noun** is a noun that shows ownership.
A **singular possessive noun** shows ownership by one person or thing.
A **plural possessive noun** shows ownership by more than one person or thing.

EXAMPLES:

Singular Possessive	Plural Possessive
These scissors belong to Lee.	The paintings done by the students are colorful.
These are Lee's scissors.	The students' paintings are colorful.

- To form the possessive of a singular noun, add an apostrophe and an **s**.
- To form the possessive of a plural noun that ends with **s**, add only an apostrophe.
- To form the possessive of a plural noun that does not end with **s**, add an apostrophe and an **s**.

It's Mine

Directions Complete each group of words by writing the possessive form of the underlined noun.

1. The ears of the elephant the ________________ ears
2. The directions from the teacher the ________________ directions
3. The names of the animals the ________________ names
4. The tail on that monkey that ________________ tail
5. The bird made by Sarah ________________ bird
6. The feathers on the ducks the ________________ feathers
7. A new animal made by Gilberto ________________ new animal
8. The wings of the butterfly the ________________ wings

Animal Analogies

Analogies are made from pairs of words that have the same relationship.

Directions **Complete the analogies using words from the box below.**

animal	bear	cattle	dragon	kitten
cold	mouse	wild	wool	turtle

1. DAISY is to PLANT as ELEPHANT is to ____________________.

2. GEESE is to GOOSE as MICE is to ____________________.

3. CAT is to DOG as ____________________ is to PUPPY.

4. ____________________ is to LAMB as the SHELL is to ____________________.

5. TAME is to ____________________ as HOT is to ____________________.

Animal Riddles

Directions **Read the riddles. Use words from the box above to fill in the blanks.**

6. What do you call a cart for carrying fire-breathing monsters?

   A ____________________ wagon

7. What do you call a war between cows?

   A ____________________ battle

8. What do you call a grizzly with no fur?

   A bare ____________________

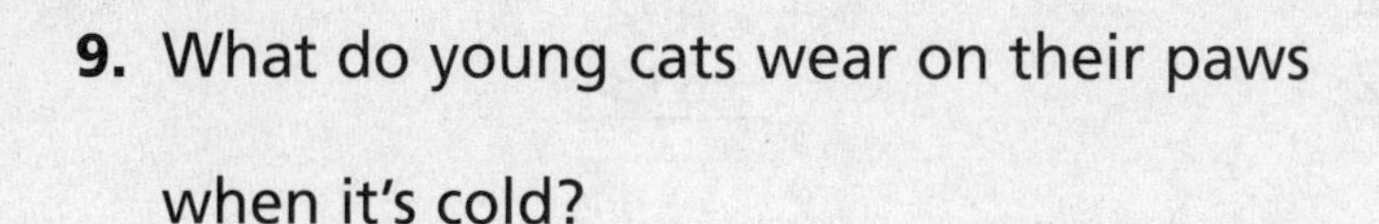

9. What do young cats wear on their paws when it's cold?

   ____________________ mittens

10. What do you call a place where furry rodents live?

    A ____________________ house

Subtracting Whole Numbers

Subtraction is often called the "opposite of addition." When you add, you put together. When you subtract, you take away. Estimating before you subtract can be helpful.

REMEMBER

To subtract, take away the bottom number from the top number.

- First subtract the ones,
- next subtract the tens,
- then subtract the hundreds.

When a bottom number is larger than a top number, regroup (borrow), and then subtract.

EXAMPLES:

	Estimate		Estimate		Estimate
237	200	479	500	\$6.93	\$7.00
− 104	− 100	− 186	− 200	−\$2.27	−\$2.00
133	100	293	300	\$4.66	\$5.00

Take It Away!

Directions Find the difference. Estimate as your first step. The first one is done for you.

		Estimate			Estimate			Estimate
1.	378 − 241 137	400 −200 200	**2.**	562 − 330		**3.**	877 − 513	
4.	329 − 117		**5.**	684 − 212		**6.**	402 − 313	
7.	\$5.99 − \$1.27		**8.**	\$7.98 − \$2.93		**9.**	\$9.50 − \$7.99	

Graphs and Charts

Directions Use the graph to answer the questions.

1. Which season had the most sunny days? ______________________

2. In which season did the number of sunny days almost equal the number of cloudy days? ______________________

3. How many cloudy days occurred during winter? ______________________

4. How many more cloudy days than sunny days occurred during winter? ______________________

5. How many more cloudy days than sunny days occurred during the school year? ______________________

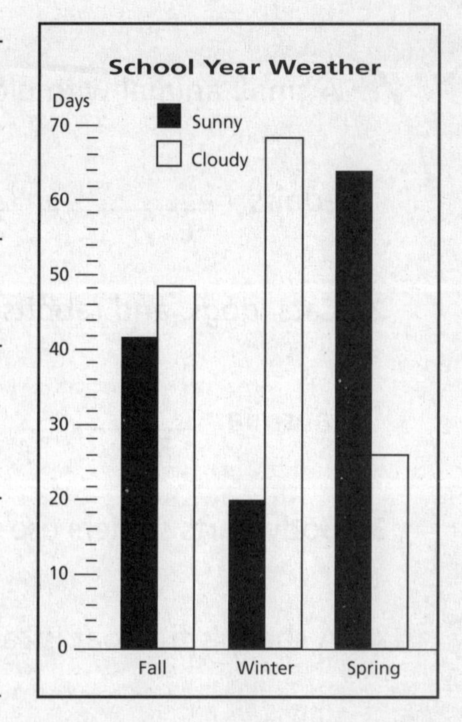

Directions Use the chart to answer the questions.

You are in charge of feeding the ducks at the pond for the next two weeks. You are given a bag of 105 crackers and you want to make sure they last.

On Day 1, you feed the ducks one cracker.
On Day 2, you feed the ducks two crackers.
On Day 3, you feed them three crackers, and so on.

Each day you feed them one more cracker than the day before.

6. a. How many crackers remain at the end of Day 4? ______

b. How many crackers remain at the end of Day 8? ______

7. On what day do you feed the ducks 10 crackers? ______

8. How many crackers can you feed the ducks on the final day? ______

End of Day	Crackers Remaining
1	104
2	102
3	99
4	
5	
6	
7	
8	
9	
10	
11	
12	
13	
14	

Animal Word Scramble

Directions **Read each definition. Then unscramble the letters to find the answers. Use the numbered letters to complete the riddle.**

1. A small animal with eight legs

edpirs ___ ___ ___ ___ ___ ___
(1) (2)

2. Cats, dogs, and rabbits are all these

lansmia ___ ___ ___ ___ ___ ___ ___
(3) (4)

3. Body parts spiders use to bite gansf ___ ___ ___ ___ ___
(5)

4. A sheep's fur coat is called its olow ___ ___ ___ ___
(6)

5. One kind of home for a spider leoh ___ ___ ___ ___
(7)

6. An animal that is hunted by another animal for food yerp ___ ___ ___ ___
(8) (9)

Now solve the riddle: What do you a call a flying spider?

___   ___ ___ ___ ___ ___ ___   ___ ___ ___ ___ ___ ___
3   1 8 4 2 7 9   5 6 4 2 7 9

A Popular Dog

The Labrador retriever, or Lab, is a popular breed of dog. Labs come in three colors: black, yellow, and chocolate. Their bodies are tall and strong. They have short hair, so they don't shed very much.

Labs are quite playful. They are also gentle and smart, so they are perfect dogs for families. Labs are easy to train, but they need a lot of attention. They also require plenty of exercise.

Labs love water, and they are great swimmers. If you throw a stick or ball into the water, Labs will swim out to retrieve it. I think they're the best dogs of all!

Directions Circle the letter of the best answer.

1. Which of the following facts about Labs is **not** true?

A. Labs come in three colors.

B. Labs have short hair.

C. Labs have a bad temper.

D. Labs love water.

2. Why is a Lab a good dog for a family?

A. A Lab is gentle and smart.

B. A Lab likes to swim.

C. A Lab's body is tall and strong.

D. The Lab is a popular breed of dog.

3. Which two words in the passage are synonyms?

A. tall and short **B.** strong and playful

C. need and require **D.** swim and train

4. The last sentence in the passage is ______.

A. a question **B.** a statement

C. a fact **D.** an opinion

CHAPTER 3

Owen Foote, Soccer Star

"I am somebody!" Owen leaped off his bed and landed in a crouched position in front of his mirror.

He squinted his eyes and stared at his reflection. First he looked at the left profile, then the right.

He made a serious face, and then his bad-guy face. "I'm the coolest guy in the whole wide world," he sang. He moved his skinny shoulders up and down like a rock star. He shuffled his feet in time to the music.

It was Friday afternoon. Owen's bed was covered with books. There was a field guide to North American fish and dolphins, a guide on amphibians and reptiles, and one on birds. Plus there were a bunch of other books about animals.

Owen loved to read about animals. He loved to read about how fast they could run, what kinds of other animals they killed, and how ferocious they were.

All this week he had been reading a book about gorillas. He thought gorillas were the greatest. Gorillas are the largest, most powerful of all living primates, the book said. But they are very peaceful.

Gorillas live in families. Their leader is called a silverback. No one bothers the silverback. He is so strong, he could tear a leopard apart with his bare hands.

The thought made Owen shiver.

The book said all the little gorillas respected the silverback. They would do anything to be near him.

Every morning after the silverback got out of his bed, the little gorillas got in it and rolled around. Having his smell on them made them feel strong and proud, like they were somebody, the book said.

Owen knew just how they felt. He felt the same way about soccer. Just putting on his cleats made him feel ten feet tall. "I am somebody," he said again as he got ready for the game.

Owen Foote, Soccer Star

Directions Using what you have just read, answer the questions.

1. Where does the story take place?

___

2. In the beginning of the story, what does Owen like to do? Tell how you know.

___

___

3. What is a **silverback**?

___

___

4. How did the young gorillas feel about the gorilla leader? Tell how you know.

___

___

5. How does Owen feel when he plays soccer?

___

___

Nouns: Common and Proper

REMEMBER

Nouns name persons, places, or things. A **common noun** names any person, place, or thing. A **proper noun** names a particular person, place, or thing.

Common nouns do not begin with a capital letter, unless they are the first word in a sentence.

EXAMPLES:

The words person, uncle, and country are all common nouns.

Proper nouns begin with a capital letter. If a proper noun has more than one word, capitalize the first letter of each important word.

The words Raul, Uncle Paul, and United States of America are all proper nouns.

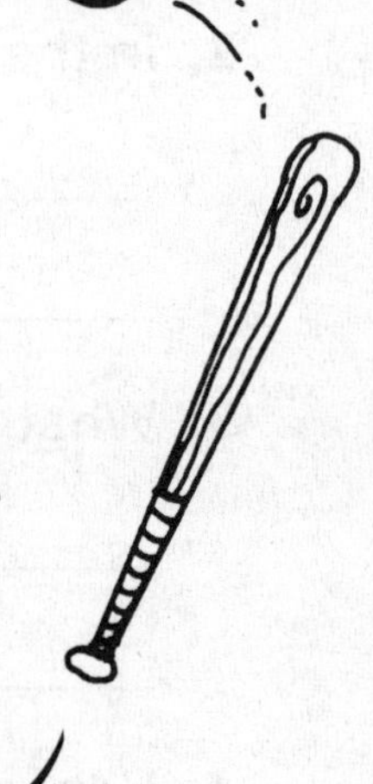

Baseball With a Capital B

Directions **Write the common nouns and proper nouns in the correct column next to each sentence. Begin each proper noun with a capital letter. There may be more than one common or proper noun in each sentence.**

	Common Nouns	Proper Nouns
1. I traveled with my baseball team last july.	____________	____________
2. The bus picked us up on bailey street.	____________	____________
3. We were going to a big game in texas.	____________	____________
4. On the first day of the trip, the bus drove through wyoming.	____________	____________
5. We saw the mountains in colorado.	____________	____________
6. My friend manuel had never been there.	____________	____________
7. The game took place at joe hollings field.	____________	____________
8. Next thursday the team is going to watch a soccer game.	____________	____________

A Baseball Trip

Directions Use the words from the box below to fill in the blanks.

remember	hurt	visit	throw

1. Next year, our family plans to ______________________ the Little League Hall of Fame.

2. While we're there, I want to try to __________________ a fast ball.

3. It would be so fast that it would _____________________ your hand if you caught it!

4. It will be a trip to _______________ !

underwater	kick	basketball	softball
swim	home runs	sports	bats

Sports for all Seasons

Directions Use the words from the box above to fill in the blanks.

My sister and I like to play sports all year round. In the winter, we practice our foul shots at the **(5)** _______________ court. When springtime comes, we pull out our mitts and **(6)** _______________. I play baseball and my sister plays **(7)** _______________. She is really good. She has hit more **(8)** _______________ than I have! In the summer, we like to **(9)** _______________ at the pool. I like to hold my breath as I swim **(10)** _______________. When the season changes, we like to **(11)** _______________ soccer balls. **(12)** _______________ can be fun all year long!

Time for Times

Multiplying can be thought of as a quick way of adding. Multiplying can save a lot of time. If you want to learn to multiply quickly, it helps to learn the multiplication tables.

REMEMBER

To multiply by one digit, multiply the top number one digit at a time.

- First multiply the ones digit,
- next multiply the tens digit,
- then multiply the hundreds digit.

When a product is larger than 9, regroup (or carry) and continue multiplying.

EXAMPLES:

	Estimate		Estimate		Estimate
314	300	78	80	\$3.19	\$3.00
× 2	× 2	× 9	× 9	× 6	× 6
628	600	702	720	\$19.14	\$18.00

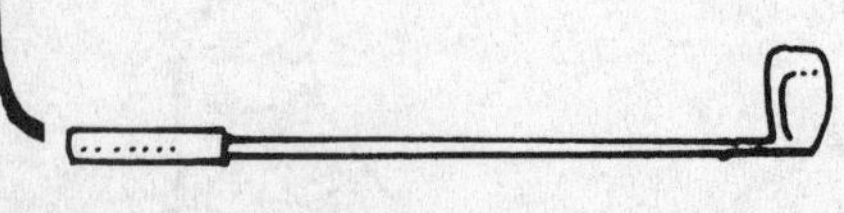

Work It Out!

Directions **Find each product. Estimate the top number as your first step. The first one is done for you.**

	Estimate		Estimate		Estimate
1. 53	50	**2.** 32		**3.** 61	
× 3	× 3	× 4		× 5	
159	150				

4. \$2.24 × 2

5. \$3.02 × 4

6. \$8.34 × 5

7. 114 × 7

8. 281 × 5

9. 527 × 8

Swimming Laps

Directions **Read the word problems and multiply to solve. Show your work.**

Lucia has joined the swim team. At practice, Lucia can swim 4 laps in 10 minutes.

1. How many laps will Lucia swim in 20 minutes? __________

Show your work.

2. How many laps will Lucia swim in 1 hour? __________

Show your work.

3. How many minutes will it take Lucia to swim 16 laps? Show your work. __________

4. Lucia's coach wants her to swim 200 laps this season. If Lucia swims 4 laps a day, how many days will it take her to swim 200 laps? __________

Show your work.

The Sports Page

Directions Read each definition. Use words from the box below to complete the puzzle.

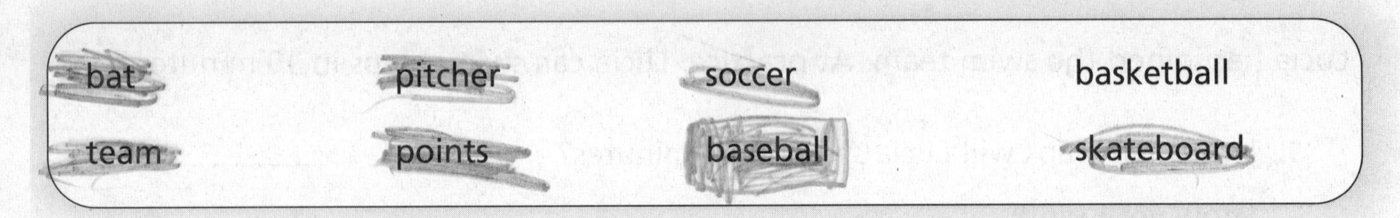

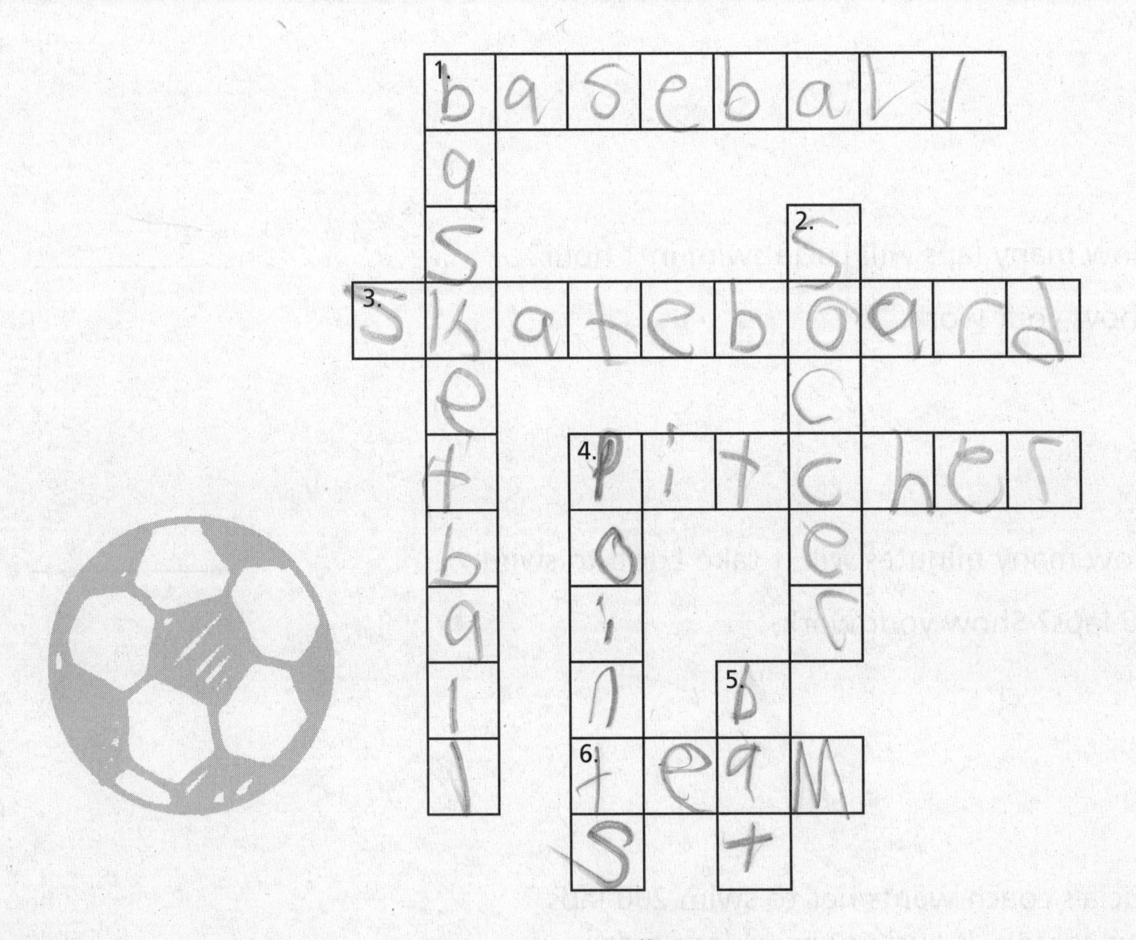

Across

1. A game played with a bat and a ball.

3. A board with wheels on it.

4. A person who throws a baseball to a batter.

6. A group of people who play a sport together.

Down

1. A game played by tossing a ball into a hoop.

2. A game played by moving a ball with a player's feet.

4. The scores of a game.

5. A long piece of wood used to strike a ball.

Estimate the Product

Directions Use rounding to estimate each product. The first one is done for you.

	Problem	Estimate		Problem	Estimate
1.	5,246 × 8	5,000 × 8 = 40,000	**2.**	545 × 6	
3.	924 × 7		**4.**	1,021 × 4	
5.	6,015 × 3		**6.**	4,258 × 2	
7.	7,924 × 5		**8.**	3,214 × 9	
9.	5,200 × 3		**10.**	8,432 × 8	
11.	5,453 × 4		**12.**	2,927 × 7	
13.	3,329 × 6		**14.**	7,438 × 2	

CHAPTER 4

How to Make a POP-UP Card

Materials Needed:

- 2 square pieces of construction paper
- pencil
- ruler
- glue
- scissors
- markers or crayons

Directions:

1. Fold one piece of construction paper in half, making a straight fold.

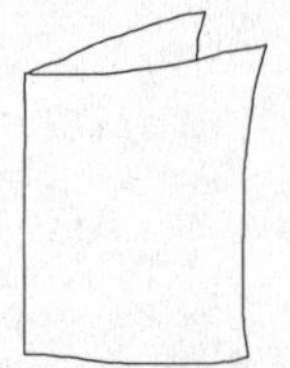

2. Keeping the paper folded, use your ruler and pencil to draw two lines near the center of the paper from the fold to the open edge. The lines should be about two inches apart.
3. Starting at the fold, cut a 2-inch slit along each line. Be careful not to cut all the way to the open edge of the paper.
4. Open the paper slightly. Push the section you have just cut out toward the open edge. It will be folded in the opposite direction from the way the rest of the paper is folded.

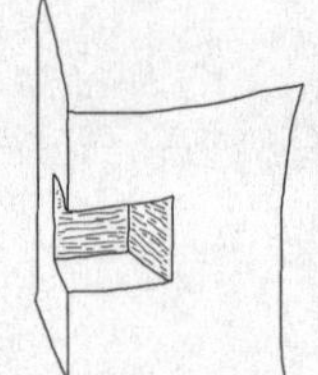

5. Open the paper all the way. The center part you cut should now stand out. Fold the pop-up part the other way so that it pops up when the card opens.

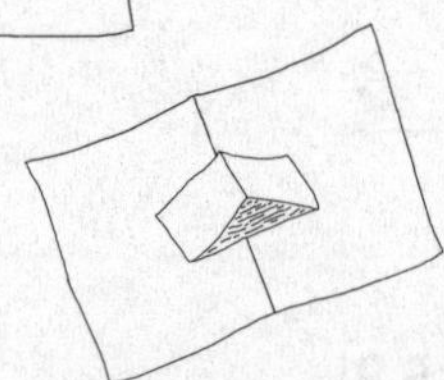

6. Glue the second piece of construction paper to the back of your pop-up card. Make sure not to get any glue on the pop-up part.
7. After the glue dries, fold the card so that the pop-up part springs out when someone opens the card. Write a message or draw a picture on it to surprise someone!
8. Use your crayons or markers to decorate the rest of the card.

How to Make a Pop-Up Card

Directions Using what you have just read, answer the questions.

1. Besides markers and crayons, what else could you use to decorate your card?

____

____

2. List 4 steps you think are the most important in making a pop-up card.

a. ____

b. ____

c. ____

d. ____

3. What are some reasons that people send cards? Name three.

____

____

____

4. Why is it important to follow the directions in order when making a pop-up card?

____

____

____

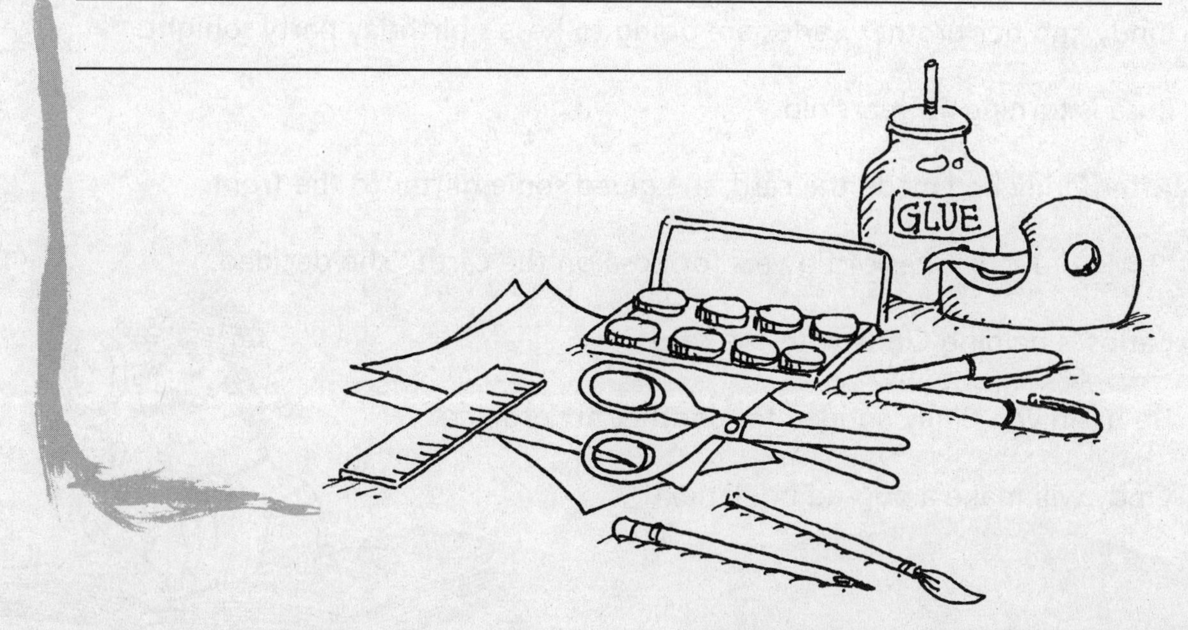

The "Write" Words

REMEMBER

Verbs show action or express a state of being.

I **wake** up. Then I **eat** breakfast.
After that, I **walk** to school. I **am** a student.

The boldface words in these sentences are **verbs**. Every sentence must have a verb. A verb can be more than one word. Often a **main verb**—the most important verb—is used with a **helping verb**. Together these words form a **verb phrase**.

EXAMPLES:

Cindy **is writing** a book for children.
She **has written** four chapters.
Laura **will help** Cindy draw pictures for the book.

The main verbs in the sentences are **writing, written,** and **help**. The helping verbs are **is, has,** and **will**. Some other common helping verbs are **am, are, was, were, have,** and **had**.

Visualize the Verbs

Directions Find the verb phrases in the sentences below. For each verb phrase, draw one line under the helping verb and two lines under the main verb.

1. Cindy was making a pop-up card for one of her friends.
2. Cindy and her brother Carlos are going to Rosa's birthday party tonight.
3. Rosa is turning 12 years old.
4. After Cindy had made the card, she glued some glitter to the front.
5. "I will draw pictures of flowers for Rosa on the card!" she decided.
6. Carlos is helping Cindy with the card.
7. He has given Cindy an idea for another art project.
8. Cindy will make a pop-up book next!

Sounds Like...

Directions **Circle the word that matches each definition.**

1. never used or worn before

knew new

2. something to write on

bored board

3. penny

sent cent

4. listened to a sound

heard herd

5. to form letters of the alphabet

write right

6. what trees are made of

would wood

A New Pal

Directions **Use the words from the list below to fill in the blanks.**

again	guess	write	because
letter	please	would	instead

Dear Paul,

Our teacher, Mr. Sanchez, has a friend who teaches school in Japan. Mr. Sanchez said the students in Japan wanted to **(7)** ____________ letters to people in America. He thought it **(8)** ____________ be a good idea for us to be their pen pals. Well, I **(9)** ____________ someone got my letter, **(10)** ____________ a few weeks later I got a message from a girl named Akira. Now I have another pen pal! **(11)** ____________ write back soon. I really want to get a **(12)** ____________ from you. I'll write to you **(13)** ____________ as soon as I find the time. Maybe I should e-mail you **(14)** ____________ !

Sincerely,

Lee

Dividing Whole Numbers

Dividing numbers is a way of asking, "How many of one number is contained in another number?" Sometimes, division results in a **remainder**—an amount that is left over.

REMEMBER

Basic division facts come from basic multiplication facts.

To solve:	You think:	You ask:	Answer:
$8\overline{)32}$	$8 \times$ _____ $= 32$	$8 \times$ what number $= 32$?	4
$7\overline{)63}$	$7 \times$ _____ $= 63$	$7 \times$ what number $= 63$?	9

EXAMPLES:

$$\begin{array}{r} 5\text{ r}4 \\ 5\overline{)29} \\ \underline{25} \\ 4 \end{array} \qquad \begin{array}{r} 56\text{ r}6 \\ 7\overline{)398} \\ \underline{35} \\ 48 \\ \underline{42} \\ 6 \end{array}$$

Dare to Divide!

Directions Find each quotient. The first one is done for you.

1. $\begin{array}{r} 7\text{ r}2 \\ 6\overline{)44} \\ \underline{42} \\ 2 \end{array}$

2. $7\overline{)53}$

3. $8\overline{)76}$

4. $2\overline{)316}$

5. $5\overline{)610}$

6. $4\overline{)637}$

7. $4\overline{)\$8.20}$

8. $3\overline{)\$8.88}$

9. $7\overline{)\$16.59}$

What Are the Chances?

Directions **Use the spinner to answer the questions.**

The pointer on the spinner below can land either in a striped or a dotted section. The pointer will not stop on a line.

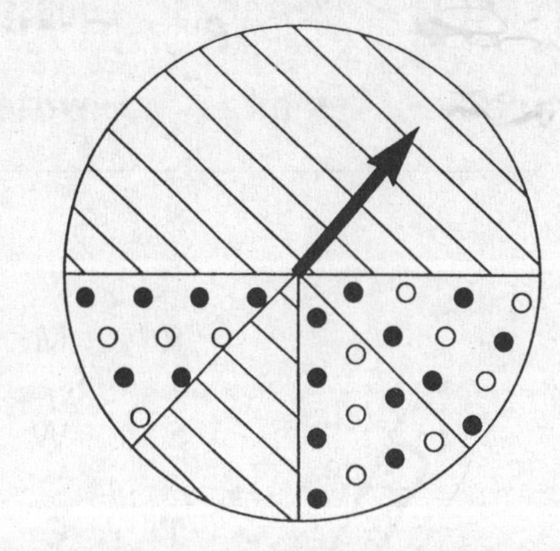

1. How many sections are on the spinner?

striped sections: ________

dotted sections: ________

total sections: ________

2. In a single spin, on what section will the pointer most likely stop? ____________

Explain your answer.

____

____

____

3. Suppose you spin the pointer 100 times. Circle the phrase that best describes the number of times the pointer is likely to land on a striped section.

fewer than 50 **exactly 50** **more than 50**

Explain your answer.

____

____

Writing Word Search

Directions Circle the words listed in the box below. The words can be found across, down, or diagonally.

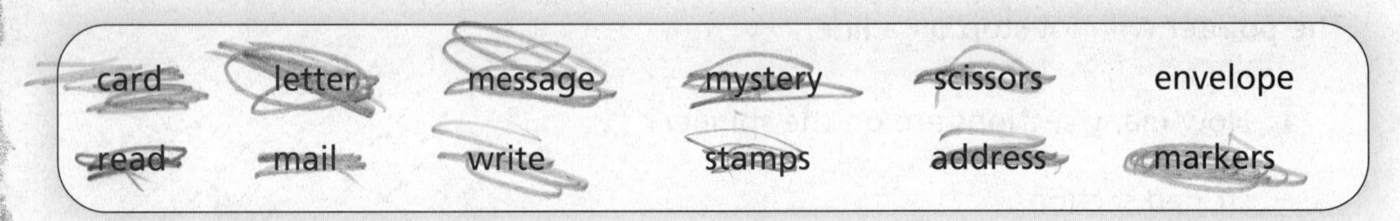

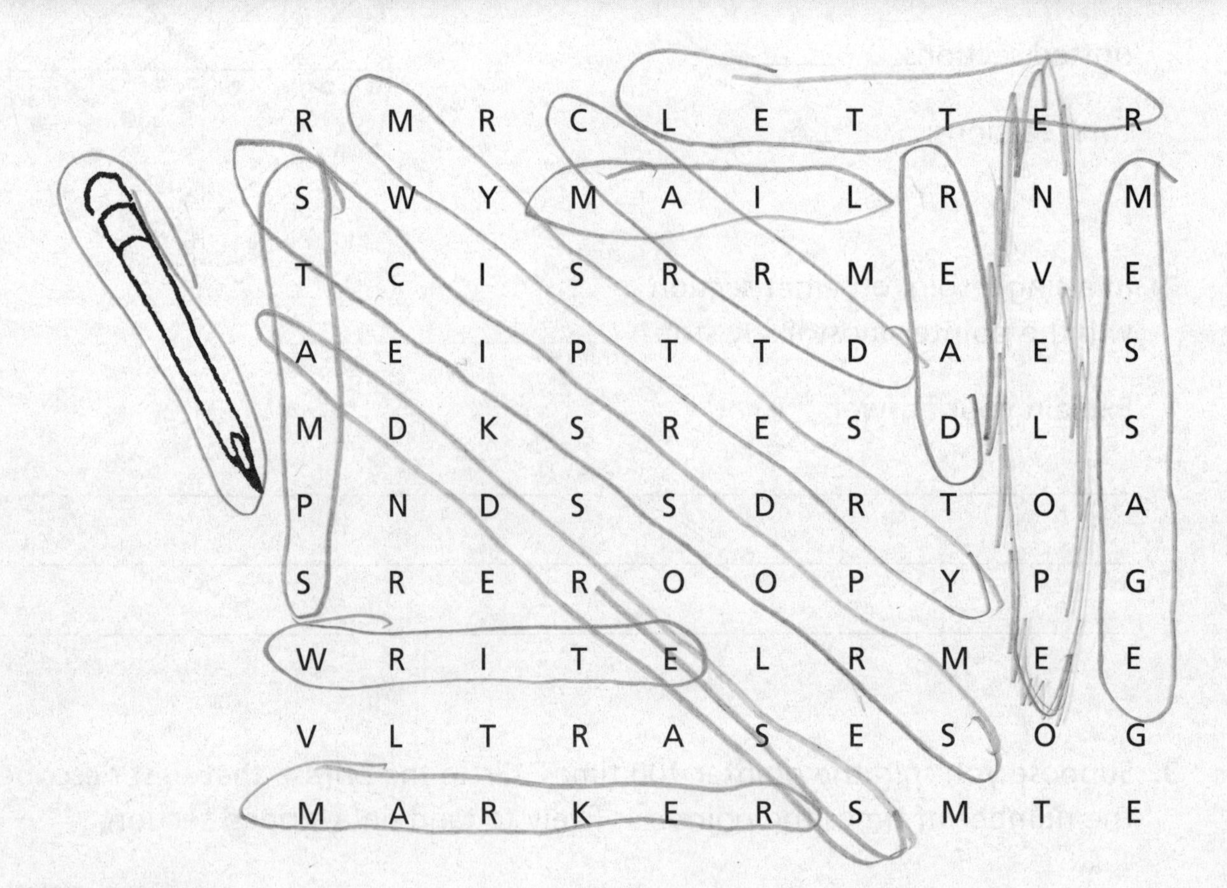

R	M	R	C	L	E	T	T	E	R
S	W	Y	M	A	I	L	R	N	M
T	C	I	S	R	R	M	E	V	E
A	E	I	P	T	T	D	A	E	S
M	D	K	S	R	E	S	D	L	S
P	N	D	S	S	D	R	T	O	A
S	R	E	R	O	O	P	Y	P	G
W	R	I	T	E	L	R	M	E	E
V	L	T	R	A	S	E	S	O	G
M	A	R	K	E	R	S	M	T	E

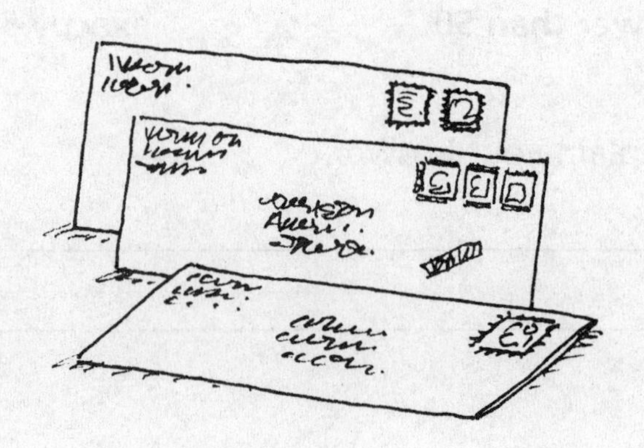

Find the Quotient

Directions Find each quotient.

1. $7\overline{)2,275}$
2. $2\overline{)2,490}$
3. $5\overline{)5,825}$
4. $9\overline{)2,763}$
5. $4\overline{)3,928}$
6. $3\overline{)9,432}$

What's the Probability?

Directions Write the answers for questions 7–9.

Julie put 11 marbles in a bag: 3 are red, 6 are blue, and 2 are green.

7. If Julie closes her eyes and draws a marble from the bag, what color marble is she most likely to draw?

____

Julie draws one marble, puts it back in the bag, and draws again. She repeats this 11 times. Use this information to answer questions 8 and 9.

8. Predict the number of times Julie will draw a red marble.

____

9. Julie draws a red marble 5 times. How does this compare to your prediction?

____

____

CHAPTER 5

Camp Gullywash

Every year in June, the normally quiet town of Gullywash comes to life. Hundreds of 11-year-old campers from around the United States come to Camp Gullywash. Camp Gullywash has everything! There is sailing, swimming, fishing, hiking, storytelling, and even rollerblading. The camp is located between a mountain and a lake, and it is surrounded by woods full of wild animals.

Every camper is told to be careful in the woods and told what to do if they see a bear or a snake. Sometimes the animals come into the camp late at night. Last year, someone left a jar of peanut butter on a table outside one of the cabins, and a bear came to try it. Bears love peanut butter even though it sticks to the roof of their mouths. This bear had such a hard time with the peanut butter getting stuck, that he sat for two hours trying to lick the inside of his mouth clean.

When the campers aren't watching bears from the safety of their cabins, they are playing sports, working on special projects, or having quiet time to do what they most enjoy.

Campers spend two weeks at the camp. On the first day, each camper fills out a schedule of what they want to do during their two-week stay. Everyone's schedule is different. For example, Susan chose to play volleyball for two hours twice a day, three days each week, and ride horses three hours a day, two days a week. Along with their major activities, campers can sign up for special trips and games. Look at the brochure to get a better idea of all there is to do at Camp Gullywash.

Camp Gullywash

Come join the fun at Camp Gullywash!

Campers can choose from the following activities:

Camp Gullywash

Directions **Using what you just read, answer the questions. Fill in the bubble next to the correct answer.**

1. Which of these statements is an opinion?

Ⓐ Camp Gullywash is near a mountain.

Ⓑ Camp Gullywash is a fun place to go.

Ⓒ Camp Gullywash has cabins.

Ⓓ Camp Gullywash has wild animals that visit.

2. Which of these statements is a fact?

Ⓐ Campers stay for two weeks.

Ⓑ It is hard to make a schedule.

Ⓒ Meals are served anytime.

Ⓓ Bears come every night.

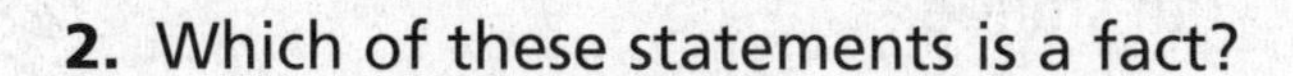

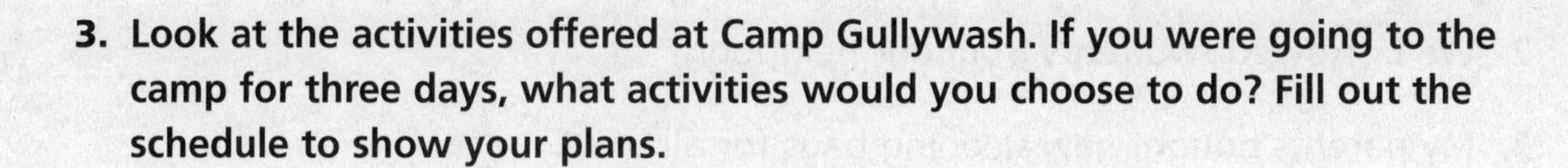

3. Look at the activities offered at Camp Gullywash. If you were going to the camp for three days, what activities would you choose to do? Fill out the schedule to show your plans.

	Day 1	Day 2	Day 3
1st Hour	Breakfast	Breakfast	Breakfast
2nd Hour			
3rd Hour			
4th Hour	Lunch	Lunch	Lunch
5th Hour			
6th Hour			
7th Hour			
8th Hour	Dinner	Dinner	Dinner

Verbs: Past Tense

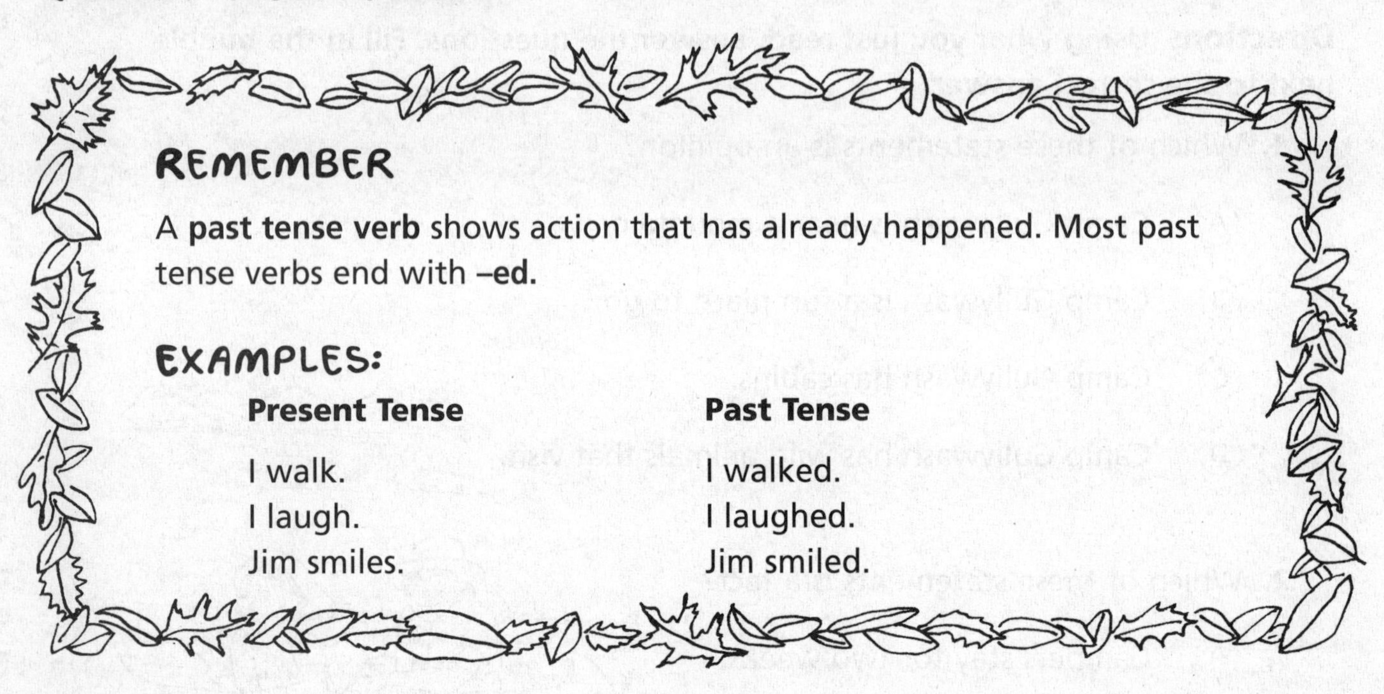

REMEMBER

A **past tense verb** shows action that has already happened. Most past tense verbs end with **–ed.**

EXAMPLES:

Present Tense	**Past Tense**
I walk.	I walked.
I laugh.	I laughed.
Jim smiles.	Jim smiled.

In the Past

Directions **Draw a line under each past tense verb in the sentences below. Some sentences may have more than one verb in the past tense.**

1. Our family went on a camping trip last spring.

2. We borrowed two tents from our neighbor.

3. My parents bought new sleeping bags for all of us.

4. We decided to camp in a wilderness park in Missouri.

5. I wanted to take our dog Racer with us.

6. I knew that Racer would love being outdoors.

7. We spent all day packing our gear.

8. My sister and I filled a bag with snacks to eat in the car.

9. Dad packed plenty of food for us to cook at the campsite.

10. On Saturday morning, we loaded the car and headed for the great outdoors.

Places to Go

Directions **Cross out the word in each group that does not belong. Write a sentence telling why you chose that word.**

1. ocean river desert ____________

____

2. cave snow skiing ____________

____

3. beach ocean cactus ____________

____

cave	forest	cactus	mountain
visitors	streets	river	beach

Not Just for Campers

California is a state that has something for everyone. **(4)** ____________ can stay in a tiny local village in the country, or in any one of the big cities. They can go from shopping along the busy city **(5)** ____________ to building sand castles on a **(6)** ____________ beside the deep, blue ocean. Just a few hours away, visitors can see a prickly **(7)** ____________ in the desert, camp under trees in the **(8)** ____________, or ski down a tall, snow-capped **(9)** ____________. Some visitors may want to take a raft on the **(10)** ____________ and ride the rapids, or visit bats inside a dark **(11)** ____________. California is a good vacation for people who like to do lots of different activities.

Working with Fractions

A **fraction** stands for a part of a whole. You probably use fractions every day, even when you talk!

"It's one-half hour until lunch."
"I've got a quarter in my pocket."
"Beth lives five-tenths of a mile from school."

REMEMBER

A fraction is written as a top number over a bottom number.

- The top number, or **numerator**, tells the number of equal parts being described.
- The bottom number, or **denominator**, tells the number of equal parts the whole is divided into.

$$\frac{\textbf{Numerator}}{\textbf{Denominator}}$$

The whole may be a single object. What fraction of the pizza is left?

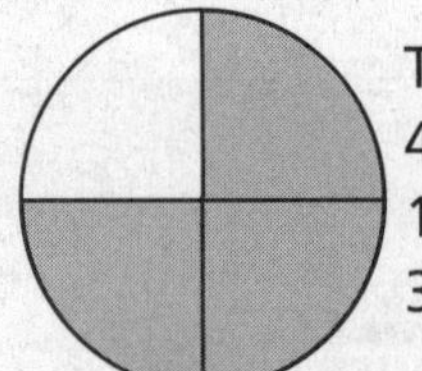

The pizza is divided into 4 equal parts. 1 part has been eaten, 3 parts are left.

$\frac{3}{4}$ $\frac{\text{Numerator}}{\text{Denominator}}$

The whole may be a group of objects. What fraction of the circles is shaded?

  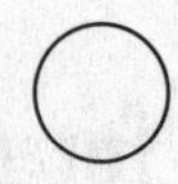  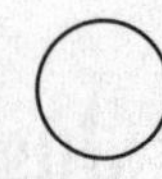

3 out of 5 circles are shaded.

$\frac{3}{5}$ $\frac{\text{Numerator}}{\text{Denominator}}$

Fraction Fun

Directions Answer the questions.

1. Complete each sentence.

a. A fraction stands for ________________ of a ________________.

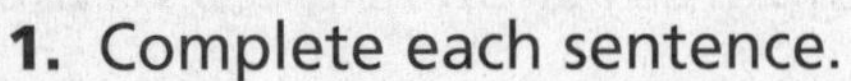

b. A whole may be a ________________ or a ________________.

2. Complete each definition.

a. numerator: The number of ________________________________.

b. denominator: The number of ____________________________.

More Fraction Fun

1. Write the correct fraction for each shaded part.

	Whole object	Shaded Part	Practice
a.		one-half	1 of 2 parts is called ______
b.		one-third	1 of 3 parts is called ______
c.		one-fourth	1 of 4 parts is called ______
d.		one-fifth	1 of 5 parts is called ______

A Fraction Frenzy

Directions **Write the fraction and the word name for the shaded part of each whole.**

2.

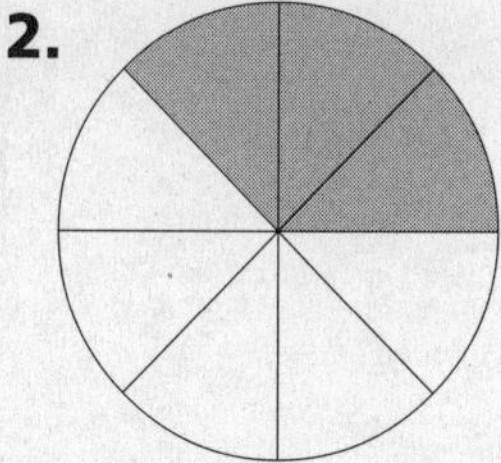

______ or ________________
fraction word name

3.

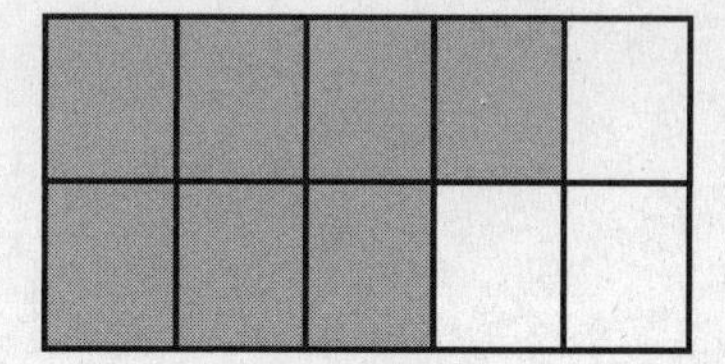

______ or ________________
fraction word name

4.

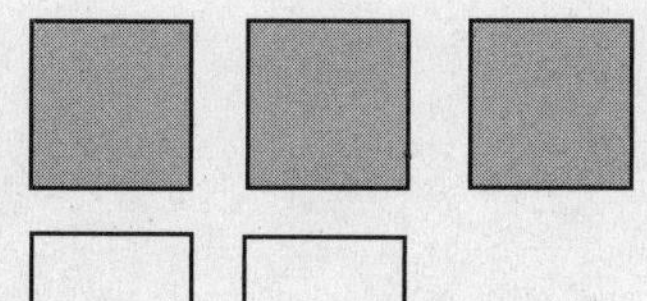

______ or ________________
fraction word name

5.

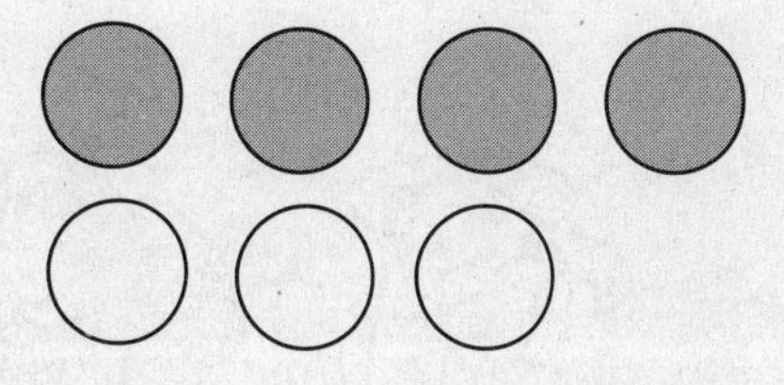

______ or ________________
fraction word name

Come To My Camp

Directions Review the article about Camp Gullywash. Pick out what you think are the most important features of the camp, such as activities or location. Make an advertisement about the camp. Be sure to include both art and writing in your ad.

Hot Dogs and Toasted Marshmallows

When you go camping, there is a lot to do during the day. But what do you do at night? Sit around a campfire, of course! You can talk, sing songs, and cook snacks over the fire. For example, you can roast hot dogs on sticks. You need a long stick that you put through the hot dogs. Then you hold the stick over the fire. You should do this only if an adult is helping you. After you eat your hot dogs, you can use the stick to roast marshmallows.

Directions **Circle the letter of the best answer.**

1. Why did the author write this passage?

A. to describe good places for camping

B. to tell how to cook treats over a campfire

C. to describe how to build a campfire

D. to describe how to set up a tent

2. Why does the author say you need a long stick?

A. Long sticks are stronger than short sticks.

B. A short stick could burn too easily.

C. The stick has to hold both hot dogs and marshmallows.

D. If the stick is too short, your hands would get too near the fire.

3. The last sentence of the passage is ______.

A. a statement **B.** a question

C. a command **D.** an exclamation

4. Which one of the following words from the passage is **not** an adjective?

A. long **B.** sweet

C. dessert **D.** delicious

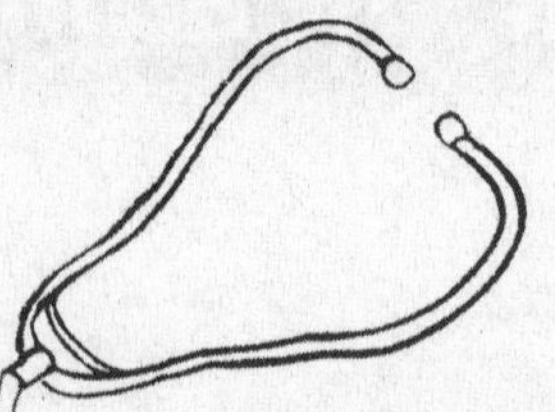

CHAPTER 6

Tools of the Trade

When you visit your doctor, he or she might look into your ears and eyes, poke your knees and elbows, and look down your throat. Then your doctor may pull out a funny-looking thing with tubes and earpieces, and a flat-looking part to put on your chest. That funny-looking thing is a stethoscope, and it's an important tool. When doctors started using stethoscopes, medicine changed for the better.

Long ago, people thought medicine was an art. That's because doctors had to be creative thinkers. Doctors asked questions to find out how the patient felt. They looked at their symptoms. They had to picture in their minds what might be wrong. Today, doctors have many tools they use to figure out what is wrong. That includes the stethoscope.

Doctors use stethoscopes to listen to how the heart and lungs work. Listening to these organs is very important. When organs sound just right, they are healthy. But sometimes the heartbeat is not quite right. Then the stethoscope helps. From the kind of sound the heart makes, the doctor figures out what illness the person may have. One kind of sound means one kind of illness; another kind of sound means another kind of illness.

The same is true of the lungs. As a person breathes in and out, the doctor listens for a healthy sound. Sometimes the lungs gurgle, as if water is inside them. The person may have a hard time breathing, or may be coughing. The stethoscope can tell the doctor what may be wrong with the lungs by the sound they make when you breathe.

You can hear heart and lung sounds for yourself. Put your ear to someone's back. You'll hear a beating heart or breathing lungs. If you had a stethoscope, you could hear those sounds much better!

Tools of the Trade

Directions Using what you have just read, answer the questions.

1. What is a **stethoscope**?

____

____

2. How did the stethoscope change medicine?

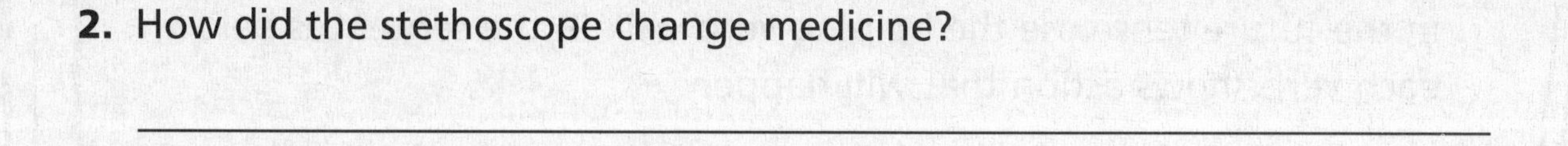

____

____

3. Name two organs listed in this article.

____

____

4. Why did doctors long ago have to be creative thinkers?

____

____

5. Is this selection real or make-believe? List three clues that help you know.

____

____

____

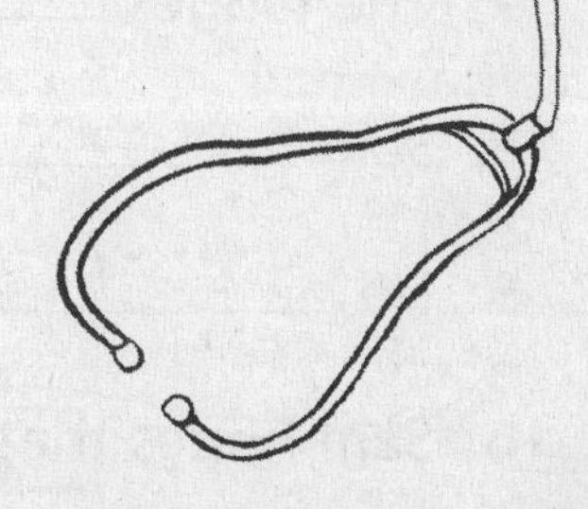

Verbs: Future Tense

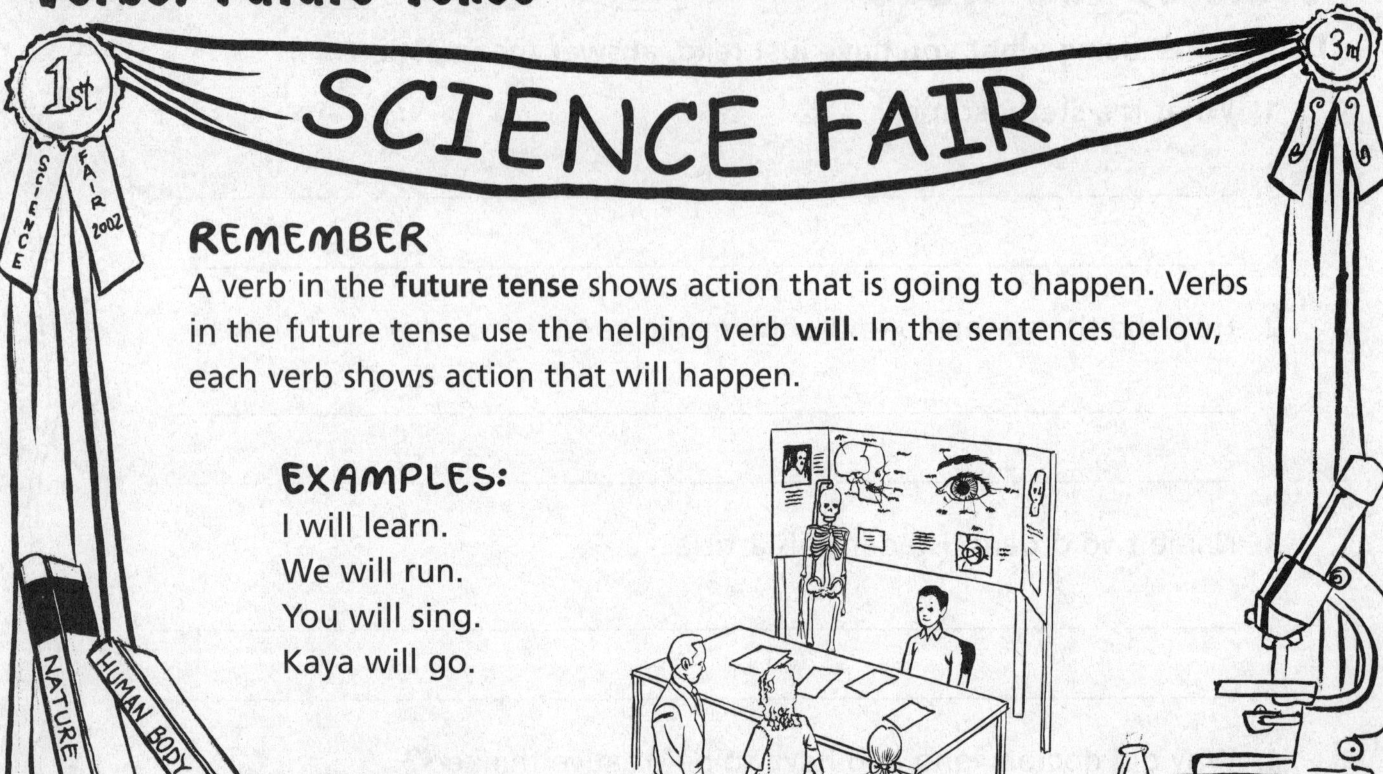

REMEMBER

A verb in the **future tense** shows action that is going to happen. Verbs in the future tense use the helping verb **will**. In the sentences below, each verb shows action that will happen.

EXAMPLES:

I will learn.
We will run.
You will sing.
Kaya will go.

A Healthy Future

Directions **Write the future tense of the verbs in parentheses to complete each sentence.**

1. Fairview School ________________ a science fair next month. (conduct)
2. Sam has no idea what topic he ________________ for his project. (choose)
3. Ms. Soto ________________ students a list of topics and suggestions. (give)
4. They ________________ three weeks to complete the project. (have)
5. Students ________________ their projects in the auditorium. (exhibit)
6. A group of judges ________________ on the three best projects. (decide)
7. The judges ________________ first-place and second-place ribbons. (award)
8. Sam ________________ the library to get ideas for a project. (visit)
9. He ________________ about experiments with the five senses. (read)
10. Sam hopes that his project ________________ a prize at the fair. (win)

Body-Talk Analogies

Directions Analogies are made from pairs of words that have the same relationship. Complete the analogies using words from the box below.

ears	wrist	knuckle	head

1. LEG is to KNEE as FINGER is to ________________.
2. CART is to HEART as LIST is to ________________.
3. SEE is to EYES as HEAR is to ________________.
4. SKULL is to SKELETON as HAIR is to ________________.

Body Blanks

Directions Read the article. Use the words from the box below to fill in the blanks.

skull	face	skeleton	bones
hair	mouth	teeth	eyes

Underneath our skin is a **(5)** ________________. It is a strong framework of **(6)** ________________ that fit together like puzzle pieces. The **(7)** ________________ makes up the bones of the head. The skull protects the brain and is also a place for **(8)** ________________ to grow. The skull also makes the structure for the **(9)** ________________, which is different for each person. All human faces have the same parts. Each has **(10)** ________________ for seeing, a nose for smelling, and a **(11)** ________________ full of strong **(12)** ________________ for eating and talking. But even though all faces have the same parts, no two faces are alike—even if they belong to twins.

Fractions: Proper and Improper

REMEMBER

Not every fraction is the same size, and large fractions have different names than small fractions.

In a **proper fraction**, the numerator is smaller than the denominator.

EXAMPLES:

$\frac{1}{2}$ $\frac{3}{4}$

In an **improper fraction**, the numerator is either: equal to the denominator (fraction equals 1), or larger than the denominator (fraction is larger than 1).

An improper fraction can be written as a **mixed number**.
A **mixed number** is a whole number and a fraction.

$\frac{3}{2} = 1\frac{1}{2}$ $\frac{7}{4} = 1\frac{3}{4}$

Putting Fractions into Action

Directions Complete the page.

1. Give an example of each.

Proper fraction ______ Improper fraction ______ Mixed number ______

2. Circle each fraction that has a value of less than 1.

$\frac{1}{8}$ $1\frac{1}{4}$ $\frac{3}{3}$ $\frac{11}{16}$ $1\frac{5}{8}$ $\frac{5}{5}$ $\frac{3}{4}$

3. Write each improper fraction as a mixed number.

$\frac{3}{2}$ = ______ $\frac{5}{4}$ = ______ $\frac{12}{8}$ = ______ $\frac{19}{16}$ = ______

4. Write each mixed number as an improper fraction.

$1\frac{2}{3}$ ______ $1\frac{4}{5}$ ______ $2\frac{3}{8}$ ______ $3\frac{6}{7}$ ______

Fraction Food Fun

Directions **Read the word problems to solve each question.**

1. Homer has 36 pretzels to share with his 5 friends. How can Homer and his friends share these pretzels equally?

____

____

2. Before Homer has a chance to give out the pretzels, two more friends arrive. How can Homer and his friends share the pretzels now?

____

____

3. The pizza below needs to be cut into 16 equal pieces. Draw lines on the pizza to show how to cut it.

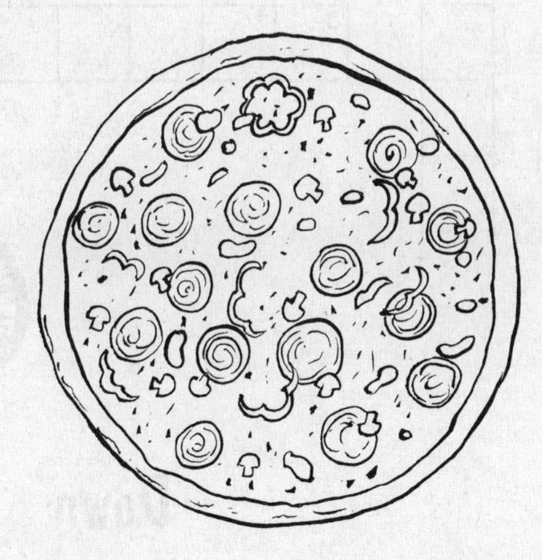

4. If Emil ate $\frac{1}{8}$ of the pizza, how many pieces did he eat?

____

5. Carlos took home $\frac{1}{4}$ of the pizza. How many pieces did Carlos take home?

____

Healthy Fun

Directions Read each definition. Use words from the chapter to complete the puzzle.

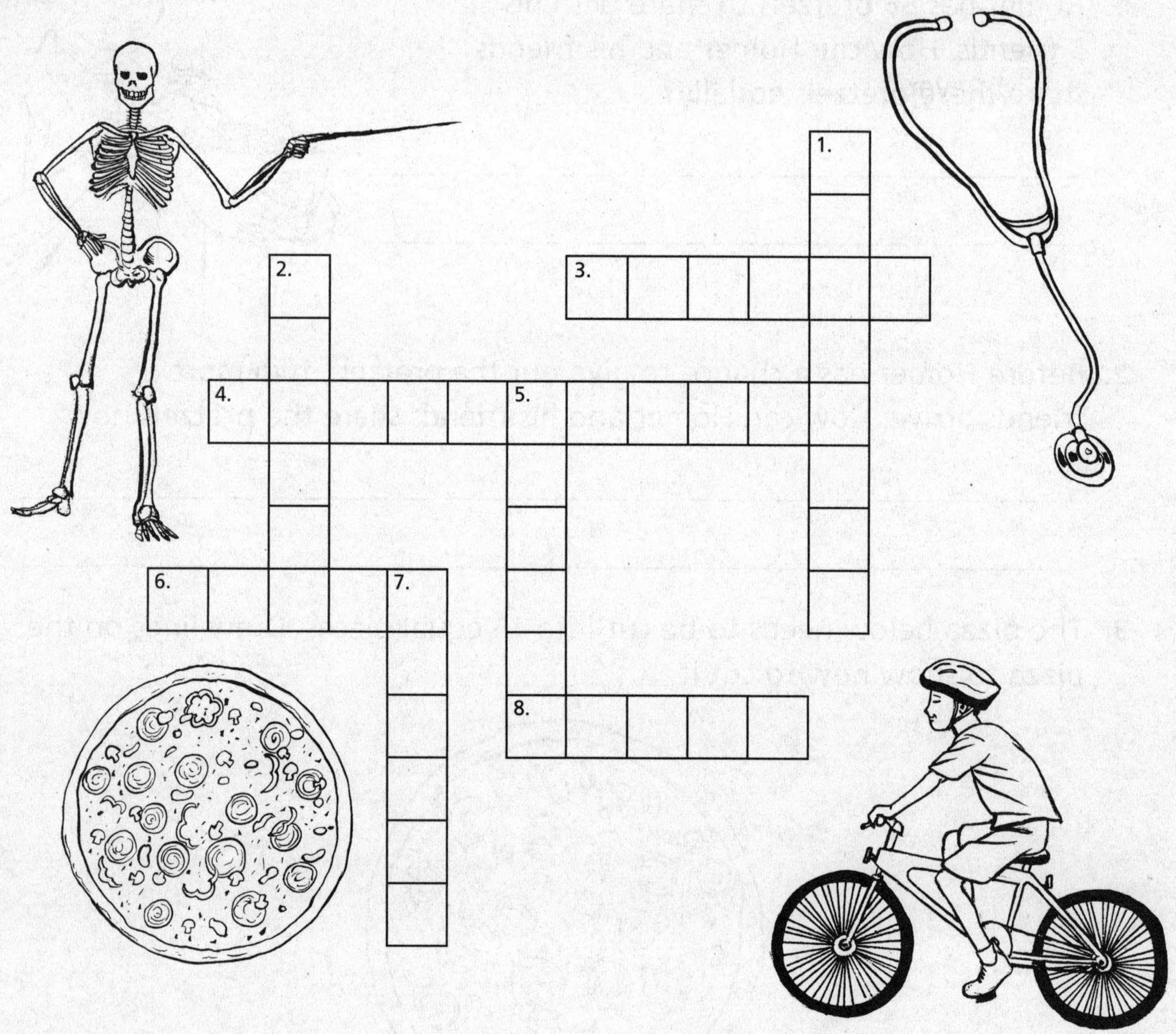

Across

3. A fraction in which the numerator is smaller than the denominator.

4. A tool used by doctors to listen to the heart.

6. Parts of the body used to chew food.

8. The part of the skeleton that protects the brain.

Down

1. All the bones in the body.

2. A verb tense that uses the word **will**.

5. The lungs and heart are examples of these body parts.

7. The study of the body and its care.

Sore Throats

Almost everyone gets a sore throat at some time, especially during the winter months. Most sore throats will appear suddenly and last between one and three days. Sore throats usually heal on their own. However, if a sore throat is very bad or lasts longer than a few days, a visit to the doctor is a good idea.

Most sore throats are caused by cold and flu viruses. Sore throats may be contagious, which means they can be passed along from person to person. Most of the time, they are passed along by people who are coughing and sneezing.

What can you do if your throat feels sore? Drink plenty of fluids; gargle with warm salt water; drink hot liquids, such as tea or chicken soup; and get plenty of rest. Also, try not to talk too much.

Directions **Circle the letter of the best answer.**

1. The purpose of this passage is to ______.

A. entertain **B.** inform

C. persuade **D.** describe a person

2. What causes most sore throats?

A. cold and flu viruses **B.** coughs and fever

C. allergies **D.** polluted air

3. Which of the following would **not** make a sore throat feel better?

A. Drinking fluids **B.** Gargling

C. Talking a lot **D.** Resting

4. Which of the following words has the opposite meaning of usually?

A. especially **B.** suddenly

C. however **D.** seldom

CHAPTER 7

SYDNEY'S PIZZA PALACE

2134 River Street (next to the movie theater)

MONDAY
2 FREE drinks when you buy a medium pizza

TUESDAY
All you can eat pizza $4.99

WEDNESDAY
Kids under 5 eat FREE

THURSDAY
Buy one, get next smaller size for half price

FRIDAY
FREE salad with purchase of a small pizza

- Choose from over 40 toppings
- Made to order
- Salad Bar
- Big Screen TV

Pizza! Pizza!

Pizza! Pizza!

Directions Using what you have just read, answer the questions.

1. Which pizza place would you go to? Why?

2. Sydney's Pizza Palace offers free meals for kids under 5 years old. What other offers might make people want to go to Sydney's Pizza Palace? Why?

3. List what you like and do not like about each ad.

Sydney's Pizza Palace

What I Like

What I Do Not Like

Danny's Pizza Parlor

What I Like

What I Do Not Like

Cooking Up Pronouns

REMEMBER

A **pronoun** can take the place of a noun in a sentence. A pronoun can be used as the subject of a sentence, or as the object.

EXAMPLE:

Tonya is baking **bread**. **She** is baking **it**.

Subject pronoun Object pronoun

The subject pronouns are **I, you, he, she, it, we, you,** and **they**.
The object pronouns are **me, you, him, her, it, us, you,** and **them**.

Pronoun Placement

Directions Write a subject pronoun or an object pronoun that could take the place of the underlined word or words in each sentence.

1. Alan wanted to make an apple pie. ________________
2. The pie would be a surprise for his mother and father. ________________
3. Alan got out the ingredients for the pie. ________________
4. Alan's sister, Lin, wanted to help Alan. ________________
5. Lin sliced the apples. ________________
6. When the pie was ready, Alan and Lin put it in the oven. ________________
7. "You and I must clean the kitchen," Alan said. ________________
8. Soon, Alan's parents came home. ________________
9. "What a nice surprise for Father and me," said Alan's mother. ________________
10. The family enjoyed eating the pie after dinner. ________________

Time to Eat

Directions Draw lines to match the definitions on the right with the words on the left.

1. dinner	**a.** a morning meal
2. breakfast	**b.** a device for cooking food
3. wheat	**c.** a meal eaten outdoors
4. picnic	**d.** food in the form of a baked loaf
5. bread	**e.** grain used in some breakfast cereals
6. stove	**f.** the main meal of the day

Good Eating

Directions Read the diary entry. Use words from the box below to fill in the blanks.

table	dinner	water	grill	salt
dishes	drink	bread	sugar	picnic

Dear Diary,

I helped my mom set the **(7)** ____________________ for a special family **(8)** ____________________ last night. First, I put out the best **(9)** ____________________. Then, I made sure everyone had a tall glass of **(10)** ____________________ to **(11)** ____________________. I set out a small bowl of **(12)** ____________________, some **(13)** ____________________ and pepper, and a large basket of freshly baked **(14)** ____________________. Mom called Dad and my two brothers to the table and we all sat down to a big, delicious meal. Sometimes, in the summer, we have a **(15)** ____________________ in the park. Dad cooks our meals on the **(16)** ____________________. Wow, that's good eating!

Adding Like Fractions

REMEMBER

Like fractions have the same bottom number (denominator). To add like fractions, add the numerators. Keep the same denominators.

Fractions can add to a sum that is:

- less than 1 < 1
- equal to 1 $= 1$
- greater than 1 > 1

EXAMPLES:

Add 2 + 1

$\frac{2}{4} + \frac{1}{4} = \frac{3}{4}$

less than one

0 1

Add 5 + 3

$\frac{5}{8} + \frac{3}{8} = \frac{8}{8}$

equal to 1

0 1

Add 4 + 3

$\frac{4}{5} + \frac{3}{5} = \frac{7}{5}$ or $1\frac{2}{5}$

greater than 1

0 1

It All Adds Up

Directions Add like fractions.

1. $\frac{2}{4} + \frac{1}{4} =$ ____ **2.** $\frac{2}{5} + \frac{2}{5} =$ ____ **3.** $\frac{4}{6} + \frac{3}{6} =$ ____ **4.** $\frac{3}{8} + \frac{5}{8} =$ ____

5. $\frac{1}{3} + \frac{1}{3}$

6. $\frac{3}{5} + \frac{1}{5}$

7. $\frac{3}{6} + \frac{2}{6}$

8. $\frac{5}{8} + \frac{4}{8}$

9. $\frac{9}{10} + \frac{4}{10}$

Ordering Pizza

Directions The prices of pizzas at several restaurants are shown on the chart below. Use the chart to answer the questions.

Pizza Prices (large sizes)			
Christy's	$10.50	Pizza House	$11.25
Luigi's	$9.75	Italian Kitchen	$10.50
Mama Mia's	$13.50	Heavenly Pizza	$12.00
King Pizza	$14.00		

1. Which two restaurants charge the same price for a large pizza?

____

2. What is the range of prices between the most expensive pizza and the least expensive pizza?

____

3. a. List the restaurants in order from most expensive to least expensive.

____

____

____

____

____

____

____

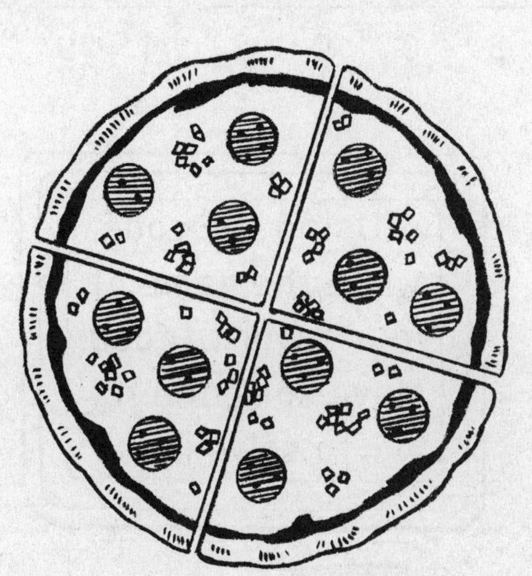

b. Whose pizza price falls in the middle?

____

Mixed-Up Menu

Directions Jill's teacher asked her to type two sentences describing her favorite food. But, Jill had her hands on the wrong letters of the keyboard! Help Jill figure out what she typed.

0 = P	1 = Q	2 = W	3 = E	4 = R	5 = T	6 = Y	7 = U	8 = I
9 = O	A = Z	D = C	E = D	F = V	G = B	H = N	I = K	J = M
O = L	Q = A	R = F	S = X	T = G	U = J	W = S	Y = H	

___ ___   ___ ___ ___ ___ ___ ___ ___ ___   ___ ___ ___ ___
J 6   R Q F 9 4 8 5 3   R 9 9 E

___ ___   ___ ___ ___ ___ ___.   ___ ___ ___
8 W   0 8 A A Q   5 Y 3

___ ___ ___ ___ ___   ___ ___ ___   ___ ___
0 8 A A Q   Y Q W   5 9

___ ___ ___ ___   ___ ___ ___ ___   ___ ___
Y Q F 3   O 9 5 W   9 R

___ ___ ___ ___ ___ ___   ___ ___ ___
D Y 3 3 W 3   Q H E

___ ___ ___ ___ ___ ___ ___ ___ ___.
0 3 0 0 3 4 9 H 8

Using the code, write the name of your favorite food. Have someone try to solve it.

____

____

____

____

____

____

A Healthy Snack

Most people like popcorn because it's so easy to make on the stove or in a microwave oven. Popcorn is also fun to eat while you watch a movie or a TV show. You can make popcorn a healthy snack by choosing a low-fat version. When you buy popcorn, be sure to read the side of the box to see how many grams of fat each serving contains.

Popcorn is quite filling because it contains a lot of fiber, which is good for you. Popcorn also contains protein and B vitamins. All of these things make popcorn an ideal snack. If you're looking for a healthy snack that can fill you up, then popcorn is the snack for you!

Directions Circle the letter of the best answer.

1. Which of the following is **not** true about popcorn?

   A. It contains fiber.

   B. It comes in a low-fat version.

   C. It contains protein.

   D. It contains sugar.

2. Why is popcorn filling?

   A. It contains meat. **B.** It comes in large bags.

   C. It contains fiber. **D.** It contains B vitamins.

3. Which of the following gives the meaning of **snack**?

   A. the main meal of the day **B.** food eaten between meals

   C. the first meal of the day **D.** a large holiday meal

4. Which adjective in the passage has the same meaning as **perfect**?

   A. easy **B.** healthy

   C. ideal **D.** good

CHAPTER 8

The Statue of Liberty

The French government gave the United States the Statue of Liberty in 1883. The statue celebrates the United States gaining freedom from England. French sculptor Frederic Auguste Bartholdie toured the United States in 1871 hoping to understand what it meant to be an American. He decided that the statue needed to be in an important, open space. On a return trip to the United States, his ships steamed into New York Harbor. After the difficult trip, the passengers cheered as they saw land. Bartholdie saw Bedloe's Island as they were making their way through the harbor and knew it would be the perfect place for his statue.

To make the statue, Bartholdie first made a frame of wood, steel, and iron. Once the frame was constructed, workers hammered copper over the frame. The statue and pedestal stand 305 feet and 1 inch tall, and were originally erected outside Bartholdie's studio in France. To get it to the United States, Bartholdie and his crew cut the statue apart, numbered each piece, and shipped it to the United States to be put together again.

He modeled the face of the statue after his mother. In one hand the statue holds a torch and in the other a tablet. "July 4, 1776," written in Roman numerals, is inscribed on the tablet. Lady Liberty, as the statue is known, wears a flowing robe and a large crown. She was made a National Monument in 1924.

Today more than 2,500 people visit the statue each day. Visitors can climb the 168 stairs to reach the top.

At the base of the statue, a poem written by Emma Lazarus is inscribed. Part of the poem reads:

Give me your tired, your poor,
Your huddled masses yearning
to breathe free,
The wretched refuse of your
teeming shore,
Send those, the homeless,
tempest-tost to me,
I lift my lamp beside the
golden door!

This poem describes the United States as a country willing to open its doors to the rest of the world. Most of the families that live in the United States came from another country originally. Lady Liberty continues to welcome new residents.

The Statue of Liberty

Directions **Using what you have just read, answer the questions.**

1. Lady Liberty wears a crown, carries a torch, and holds a tablet with "July 4, 1776" inscribed on it. Describe what each of these could mean to someone coming to the United States from another country.

Crown: ______________________________

Torch: ______________________________

Tablet: ______________________________

2. In the story, Mr. Bartholdie toured the United States to understand what it was like to be an American. If you were going to tour the USA, what cities or states would you want to visit and why? List three choices.

City or state I want to see	Why I want to see it
a. ______________________	______________________
b. ______________________	______________________
c. ______________________	______________________

3. What would be another good title for this selection?

____

4. What does the word **freedom** mean to you?

____

____

Adjectives

REMEMBER

An **adjective** is a word that describes or gives more information about a noun or a pronoun. Adjectives usually come **before** the nouns they describe.

EXAMPLE:

Our class visited a **large** museum. The director gave a **short** speech.

Predicate adjectives can come **after** the words they describe. A predicate adjective is an adjective that follows a form of the verb **be** and describes the subject of the sentence. **Am, is, are, was,** and **were** are forms of **be**.

The museum is **large**. The speech was **short**.

Artistic Adjectives

Directions **Find the predicate adjective for each underlined word in the sentences. Write the predicate adjective or adjectives next to the sentence.**

1. I was excited about our field trip to the museum. ____________
2. The museum is famous for its modern art. ____________
3. But some art in the museum is old. ____________
4. The museum guide was helpful. ____________
5. She was knowledgeable about all the art. ____________
6. I was thrilled to see so many paintings. ____________
7. And the sculptures were so graceful. ____________
8. The tour was exciting. ____________
9. But I thought it was long. ____________
10. We were happy but tired after the museum visit. ____________

All About Art

Directions Cross out the word that does not belong in each group.

1. color	music	song
2. paint	pencil	stories
3. picture	frame	music
4. books	stories	museum
5. song	paper	canvas
6. paint	sculpt	sing

Visit to the Art Museum

Directions Read the story. Use words from the box below to fill in the blanks.

art	paint	music	oil
stories	paintings	colors	pencil

Yesterday we went to the art museum. The guide told us **(7)** ____________________ about the many **(8)** ____________________ hanging on the walls. She said many artists use **(9)** ____________________ paint on linen canvas when creating a picture, while others like to make drawings using **(10)** ____________________ on paper. The museum was filled with vibrant pieces of **(11)** ____________________ everywhere. There was even a string quartet playing, filling the halls of the art museum with beautiful **(12)** ____________________. I think I'll go to the art museum more often!

Subtracting Fractions

REMEMBER

To subtract like fractions, subtract the numerators. Keep the same denominator.

- In Example 1, subtract the top numbers.
- In Example 2, write 1 as the improper fraction before subtracting.
- In Example 3, change the mixed number to an improper fraction before subtracting.

EXAMPLE 1

$\frac{7}{8} - \frac{2}{8} = \frac{5}{8}$

EXAMPLE 2

$1 = \frac{3}{3}$

$- \frac{2}{3} = \frac{2}{3}$

Write 1 as $\frac{3}{3}$
Subtract 3 − 2

EXAMPLE 3

$1\frac{1}{4} = \frac{5}{4}$

$- \frac{3}{4} = \frac{3}{4}$

Change $1\frac{1}{4}$ to
$\frac{4}{4} + \frac{1}{4} = \frac{5}{4}$
Subtract 5 − 3

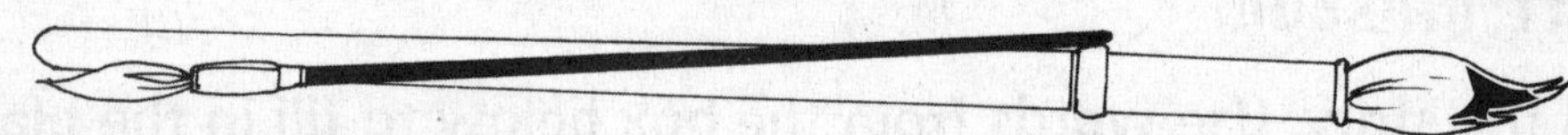

Picture Perfect Fractions

Directions Subtract the fractions. Reduce to the lowest terms, if possible.

1. $\frac{3}{4} - \frac{1}{4} =$
2. $\frac{4}{5} - \frac{2}{5} =$
3. $\frac{7}{6} - \frac{3}{6} =$
4. $\frac{9}{8} - \frac{2}{8} =$
5. $\frac{2}{3} - \frac{1}{3}$
6. $\frac{4}{5} - \frac{1}{5}$
7. $\frac{5}{6} - \frac{3}{6}$
8. $\frac{8}{8} - \frac{6}{8}$

Directions The shaded figures below show fraction subtraction. Write the fraction or mixed number below each figure.

9.

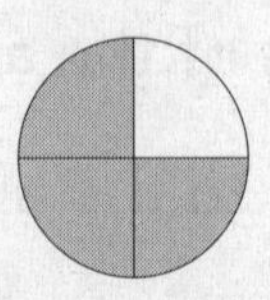 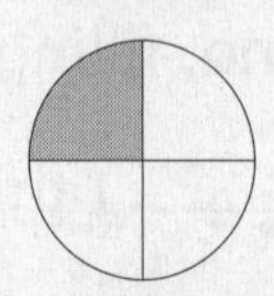 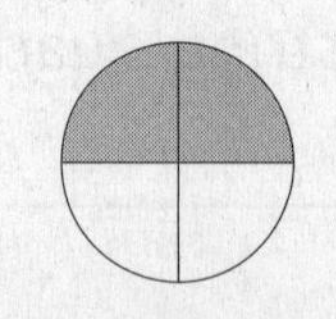

__________ − __________ = __________

10.

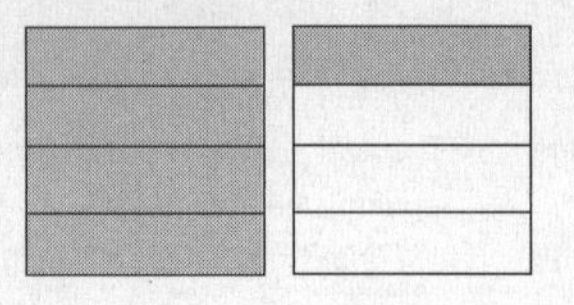 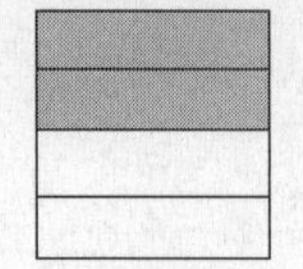 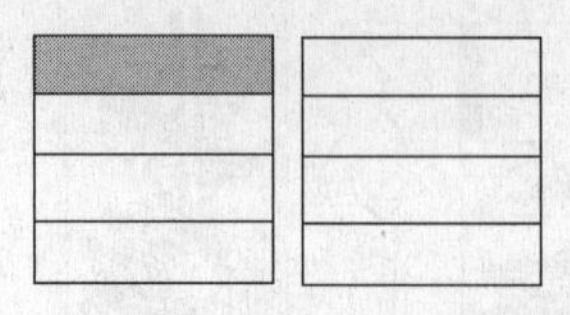

__________ − __________ = __________

Going Places

Directions The grid to the right is part of a map. Each square on the grid is 1 square block.

North

West

Sport Shop

Zoo

Bus Stop

East

South

Label each of the places below in the correct square block on the grid.

1. The library is 5 blocks north and 3 blocks east of the zoo.
2. The museum is 2 blocks south and 1 block west of the sport shop.
3. The theater is 6 blocks north and 2 blocks west of the bus stop.
4. The art store is 4 blocks south of the library.
5. The craft store is 6 blocks east of the zoo.
6. Give directions telling how to get from the bus stop to the sport shop.

____

____

7. Give directions telling how to get from the sport shop to the zoo, stopping at the theater on the way.

____

____

____

Art and You

Directions **Museums are places with all kinds of art. They have sculptures and paintings. Suppose you are an artist and you want to make a work of art. What kind of art would you like to make? Draw a picture of your art. Write some sentences describing how you made it.**

Subtracting Fractions

Directions Subtract the fractions. Reduce the answer to lowest terms, if necessary.

1. $\frac{9}{10} - \frac{3}{10} =$ ______ **2.** $\frac{3}{5} - \frac{1}{5} =$ ______

3. $\frac{5}{12} - \frac{2}{12} =$ ______ **4.** $\frac{7}{8} - \frac{3}{8} =$ ______

5. $\frac{9}{9} - \frac{5}{9} =$ ______ **6.** $\frac{4}{8} - \frac{2}{8} =$ ______

7. $\frac{5}{14} - \frac{1}{14} =$ ______ **8.** $\frac{6}{7} - \frac{5}{7} =$ ______

From Here to There

Directions Joe and Kevin live in the same neighborhood. The grid represents this neighborhood. Each square on the grid is 1 square block. Use this information to answer questions 9–11.

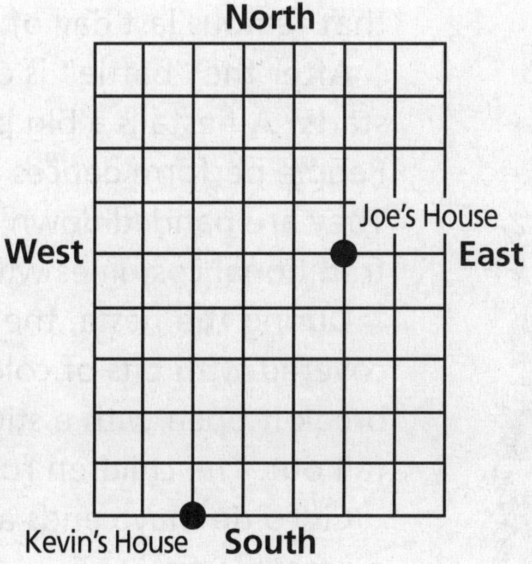

9. Kevin's house is 3 blocks west and 5 blocks south of Joe's house. Label Kevin's house on the map.

10. Give directions telling how to get from Kevin's house to Joe's house.

____

____

11. Give directions telling how to get from Joe's house to Kevin's house.

____

____

CHAPTER 9

Cinco de Mayo

Holidays are for remembering. In Mexico, and in Hispanic-American towns in the United States, May 5th has a special meaning. That's the day that a great battle took place in the city of Puebla, Mexico, in 1862. On Cinco de Mayo ("Fifth of May"), General Zaragoza, with his army of 2,000 men, finally won the battle. It was the Battle of Puebla, and it was long and horrible. For more than a week, General Zaragoza and his men fought the French army in the rain and cold. With 6,000 men, the French army was three times as big as the Mexican army. But in the end, on Cinco de Mayo, the Mexicans won the battle.

Every year since the Battle of Puebla, Cinco de Mayo is celebrated as a holiday. In Mexico, people begin the day remembering the battle. Men dress up to look like the French and Mexican armies. They act out that famous last day of the battle when General Zaragoza's army won.

After the "battle" is over, the fiesta begins. That's when the fun starts! A fiesta is a big party. Mariachi bands play guitars and drums. People perform dances in the street. Many of the dances are very old. They are handed down through families as tradition. The dancers wear traditional costumes with bright colors.

During the fiesta, the children get to hit a piñata. The piñata is covered with bits of colorful paper. The children take turns trying to break it open with a stick. When the piñata breaks, candies and toys fall out. The children rush to pick up the candies and toys.

Cinco de Mayo ends as colorful fireworks light up the night sky. It is a great fiesta.

Cinco de Mayo

Directions Using what you have just read, answer the questions.

1. What does **Cinco de Mayo** mean?

____

2. What does Cinco de Mayo celebrate?

____

____

3. Circle the sentence that is an opinion.

   That's when the fun starts!

   Many of the dances are old.

4. What do you think is the best part of Cinco de Mayo?

____

____

5. Is this selection real or make-believe? List three clues that help you know.

____

____

____

Sentence Types

HAPPY BIRTHDAY

REMEMBER

A **sentence** is a group of words that expresses a complete thought. Sentences always begin with a capital letter and end with a punctuation mark. There are four kinds of sentences.

A **statement** is a sentence that tells something. It ends with a **period**.

I made a birthday cake today.

A **question** is a sentence that asks something. It ends with a **question mark**.

Whose birthday is it?

A **command** is a sentence that tells someone to do something. It ends with a **period**.

Hand me the spoon.

An **exclamation** is a sentence that shows strong feeling. It ends with an **exclamation mark**.

What a delicious cake!

A Sentence Celebration

Directions Read each sentence. Decide whether it is a statement, question, command, or exclamation. Write the sentence type on the line.

1. What kind of cake should I make? ____________________

2. Maybe I'll make a lemon cake. ____________________

3. Hand me the red cookbook on the shelf. ____________________

4. Do I have all I need to make the cake? ____________________

5. We're out of eggs! ____________________

Compound Words

Directions Make compound words by drawing lines to connect a word from the left column with a word from the right column. Write the new word on the line. The first one is done for you.

1. after — one — afternoon
2. every — out — ____________
3. may — noon — ____________
4. with — be — ____________

Skydiving at Sixty

Directions Read the diary entry. Use words from the box below to fill in the blanks.

airplane	birthday	breakfast
everyone	grandmother	sometime

Dear Diary,

Last Saturday was a special day. My **(5)** ____________ had her sixtieth **(6)** ____________. She gave herself a very special gift. She has always wanted to go skydiving **(7)** ____________, so she did it. Right after **(8)** ____________, we all went out to the airfield. **(9)** ____________ watched as Grandma jumped out of the **(10)** ____________ and floated safely to the ground. She said it was more fun that any roller coaster ride!

Adding Like Fractions

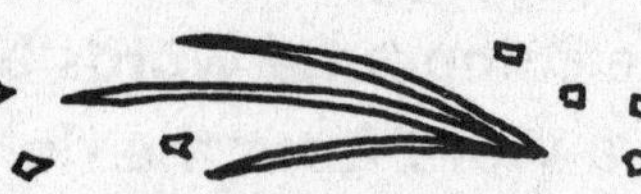

REMEMBER

To add **unlike** fractions, change them to **like** fractions.

Choose a **common denominator**—a denominator that can be used for both fractions.

Write equivalent like fractions using the common denominator.

Add the like fractions.

EXAMPLES:

In these examples, the larger denominator can be used as the common denominator.

$$\begin{array}{r} \frac{1}{2} = \frac{2}{4} \\ + \frac{1}{4} = \frac{1}{4} \\ \hline \frac{3}{4} \end{array} \qquad \begin{array}{r} \frac{3}{8} = \frac{6}{16} \\ + \frac{3}{16} = \frac{3}{16} \\ \hline \frac{9}{16} \end{array}$$

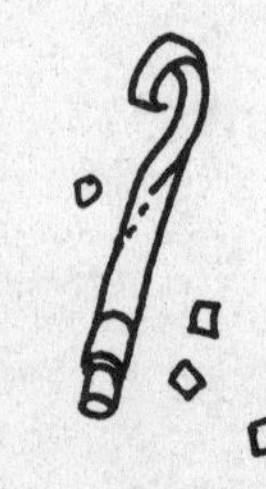

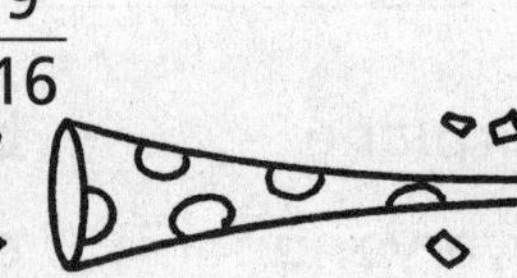

Fraction Fun

Directions Write equivalent fractions. The first one is done for you.

1. $\frac{3}{4} = \frac{\boxed{6}}{8}$
2. $\frac{1}{2} = \frac{\square}{4}$
3. $\frac{2}{3} = \frac{\square}{6}$
4. $\frac{5}{8} = \frac{\square}{16}$
5. $\begin{array}{r} \frac{1}{2} = \frac{\square}{8} \\ + \frac{1}{8} = \frac{1}{8} \\ \hline \end{array}$
6. $\begin{array}{r} \frac{3}{8} = \frac{3}{8} \\ + \frac{1}{4} = \frac{\square}{8} \\ \hline \end{array}$
7. $\begin{array}{r} \frac{2}{3} = \frac{\square}{6} \\ + \frac{1}{6} = \frac{1}{6} \\ \hline \end{array}$
8. $\begin{array}{r} \frac{1}{10} = \frac{1}{10} \\ + \frac{3}{5} = \frac{\square}{10} \\ \hline \end{array}$

Writing Fractions

Directions Write a fraction below each figure to show the addition.

9.

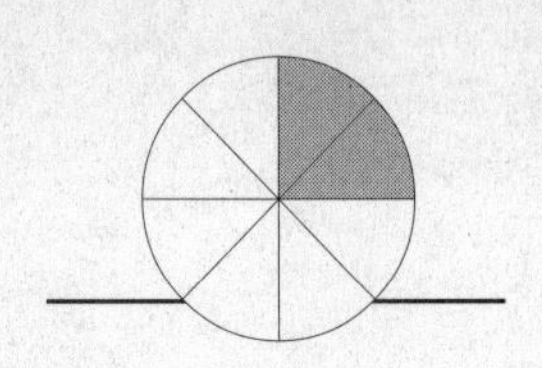
\+
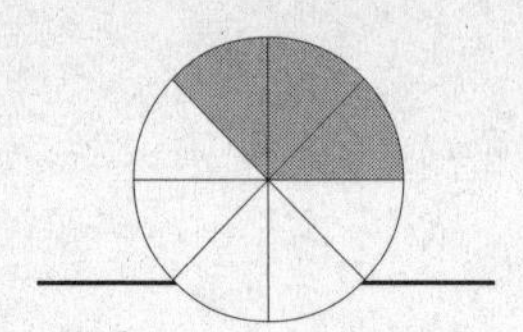
=
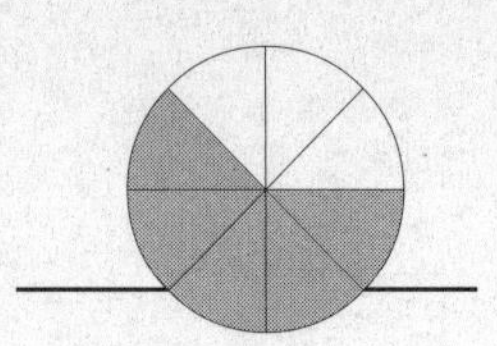

Party Fractions

Directions **Read the word problems and solve.**

1. Wendy has a 1-gallon jug of punch for Amanda's birthday party. Wendy doesn't know it, but the jug has tipped over and has started to spill out. About $\frac{1}{8}$ gallon will spill out each hour. The party is going to start in 3 hours.

a. How much punch will spill out by the time the party starts? Write your answer as a fraction of a gallon. ________

b. About how much punch will be in the jug when the party starts? Write your answer as a fraction of a gallon. ________

2. Amanda discovers the spilled jug just as the party starts. She decides to pour the remaining punch into plastic cups. Each cup holds 9 fluid ounces.

a. About how many fluid ounces of punch remain in the jug? Hint: 1 gallon equals 128 fluid ounces. ________

b. About how many cups will Amanda fill with punch? ________

Join the Celebration

3. It takes you 30 minutes to clean your room. What fraction of an hour is that? Circle your answer.

$\frac{1}{4}$ hour $\frac{2}{4}$ hour $\frac{3}{4}$ hour

4. It takes you 15 minutes to get to Amanda's house. What fraction of an hour is that? Circle your answer.

$\frac{1}{4}$ hour $\frac{2}{4}$ hour $\frac{3}{4}$ hour

5. What time should you start cleaning your room if the party starts at 3:15 P.M.? ________

Cinco de Mayo Code

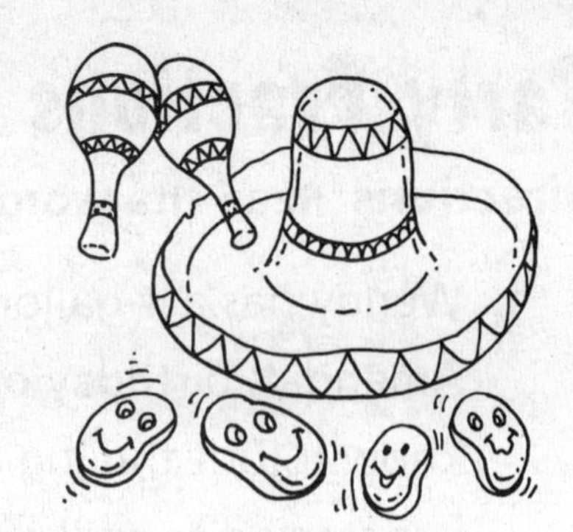

Directions **Look at the story on page 68. Solve the code using the key below. Each letter in the key goes with a symbol.**

KEY:

A	B	C	D	E	F	G	H	I	L	M	O	P	R	S	T	U	V	Y
🔔	✦	✪	💧	⚐	★	✿	◄	☺	■	❄	✧	◆	◆	×	✋	⊂	❖	○

EXAMPLE:

V I C T O R Y

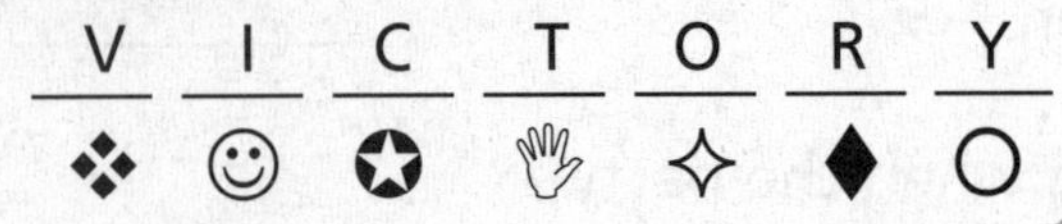

1. In Puebla, Mexico, a famous BAttLe came to an end on Cinco de Mayo.

2. In the battle, one ARMY fought another.

3. In Mexican tradition, dancers wear COLOREUL costumes.

4. People PeRFORM many traditional dances.

5. With parades, costumes, and piñatas, Cinco de Mayo is a GReat holiday.

6. Cinco de Mayo is also a musical holiday, with mariachi bands playing GUitaRS.

Fractions

Directions Write equivalent fractions.

1. $\frac{1}{2} = \frac{\square}{10}$ 2. $\frac{2}{3} = \frac{\square}{9}$ 3. $\frac{1}{2} = \frac{\square}{8}$ 4. $\frac{4}{5} = \frac{\square}{10}$

Directions Write equivalent fractions. Then, add the like fractions.

5. $\frac{1}{2} = \frac{\square}{10}$
$+ \frac{3}{10} = \frac{3}{10}$

6. $\frac{2}{3} = \frac{\square}{9}$
$+ \frac{4}{9} = \frac{4}{9}$

7. $\frac{3}{5} = \frac{\square}{10}$
$+ \frac{3}{10} = \frac{3}{10}$

Directions Add the fractions.

8. $\frac{2}{5} + \frac{8}{15}$

9. $\frac{2}{5} + \frac{3}{10}$

10. $\frac{1}{3} + \frac{1}{6}$

11. $\frac{5}{6} + \frac{1}{12}$

12. $\frac{1}{4} + \frac{3}{8}$

13. $\frac{4}{9} + \frac{1}{3}$

CHAPTER 10

Paul Bunyan

Have you heard about the mighty logger, Paul Bunyan? Well, it's time you did! Some are sure that Paul stood taller than the tallest redwood tree. They say he was stronger than 50 bears put together. And not only that, Paul was smart. But what some don't know is that Paul Bunyan had a heart of gold.

Paul had a huge helper, Babe the blue ox. One day when Paul was cutting down trees in the snow, he came across Babe. She was under a pile of snow. When Paul pulled her out, she was frozen stiff and blue as the sky. Paul blew on her three times. She warmed right up, but stayed blue. From that day on, Paul and Babe the blue ox worked together.

Back when our country was young, Paul and Babe cleared land for settlers. They cut down trees to make room for farms and towns. They had much work to do. When they first got to the Dakotas, the trees were so thick that Paul had to pry them apart. And the roads were so tangled, they had to fix them. If they didn't, the settlers would have a terrible time traveling in the new land. To straighten out the crooked road, Paul hooked the far end of the road to Babe. Then he had her pull and pull until it lay straight.

Let's get back to the trees in the Dakotas. Paul cut for days and days. He could fell a huge tree with just one swing of his ax. And Babe would help by dragging the trees away. Later, they would cut the trees into logs for the settlers to use. Paul and Babe felt good about their work. They knew that people needed a place to live. They knew they needed the logs to build homes and other buildings. So Paul cut down the trees.

Finally, he cut down the very last tree in the Dakotas. He thought he'd be glad that his work was done, but Paul felt sad. He thought about the settlers. "They won't get to cool off under a tree," he told Babe. "And they won't hear leaves rustle in the wind." Paul did some hard thinking and decided to give up logging. "We're still going to work with trees," he told Babe. "They'll just be tiny ones." With that, Paul and Babe rushed off to begin their new work.

Paul Bunyan

Directions Using what you have just read, answer the questions.

1. What is one way that Paul Bunyan showed he had a heart of gold?

2. Why did Paul Bunyan cut down trees?

3. How did Paul feel after he cut down the last tree in the Dakotas? Why did he feel this way?

4. What do you think Paul and Babe's new work will be?

5. What clues from the story did you use to answer the question?

Comparing with Adverbs

REMEMBER

Adverbs describe an action verb or a form of **to be**. They tell when, where, or how. They can come before or after the verb they describe.

EXAMPLES:

She **often** walks for exercise. Jeb left his baseball glove **upstairs**.

Adverbs can be used to make comparisons.

Roberto runs **faster** than Craig does. Carrie runs **fastest** of all the girls.

Ad –**er** or –**est** to adverbs with one syllable and a few adverbs with two syllables. Use **more** or **most** before adverbs that end in –**ly** (except early).

soon → sooner → soonest
closely → more closely → most closely

Some adverbs have special forms for comparisons.

well → better → best
little → less → least

Adverb Advantage

Directions Read the sentences below. Write the correct form of each adverb in parentheses.

1. Jon tried ____________________ than Kevin to win the race. (hard)
2. Clara swims the ____________________ of all the swimmers on the team. (strong)
3. The swim team works out ____________________ than the track team does. (often)
4. Coach Cardenas seemed to watch the red team ____________________ than he watched the blue team. (closely)
5. The crowd for the red team shouted ____________________ than the crowd for the blue team. (loudly)
6. I think Roberto touched the wall ____________________ than Eric did. (soon)

Getting to the Root of It

Directions Write the root words of the words below.

1. writing ____________________ **2.** flying ____________________

3. reading ____________________ **4.** running ____________________

5. bored ____________________ **6.** walking ____________________

7. working ____________________ **8.** swimming ____________________

So Much to Do!

Directions Read the essay. Use words from the box below to fill in the blanks.

draw	flying	reading	running
bored	study	swimming	writing

There's nothing better on a hot summer's day than **(9)** ____________________ at my neighborhood pool. Sometimes, with the nice weather, it is difficult to stay inside and **(10)** ____________________ my schoolwork. I'd rather be outside **(11)** ____________________ with my dog, or **(12)** ____________________ my model airplane. On rainy days, I stay inside and **(13)** ____________________ with my colored pencils. I also enjoy **(14)** ____________________ books and **(15)** ____________________ in my diary. There's so much to do that I'm never **(16)** ____________________.

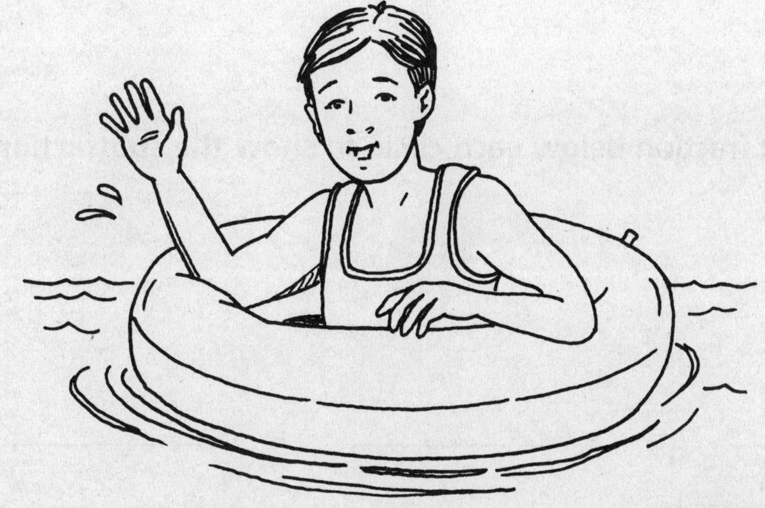

Subtracting Like Fractions

REMEMBER

To subtract **unlike** fractions change them to **like** fractions.

- Choose a common denominator.
- Write equivalent like fractions using the common denominator.
- Subtract the like fractions.

EXAMPLES:

$$\begin{array}{r} \frac{3}{4} = \frac{3}{4} \\ -\frac{1}{2} = \frac{2}{4} \\ \hline \frac{1}{4} \end{array} \qquad \begin{array}{r} \frac{7}{8} = \frac{14}{16} \\ -\frac{5}{16} = \frac{5}{16} \\ \hline \frac{9}{16} \end{array} \qquad \begin{array}{r} \frac{7}{8} = \frac{14}{16} \\ -\frac{5}{16} = \frac{5}{16} \\ \hline \frac{9}{16} \end{array}$$

Subtract It

Directions Complete each subtraction problem.

1. $\begin{array}{r} \frac{1}{2} = \frac{\square}{8} \\ -\frac{3}{8} = \frac{3}{8} \\ \hline \end{array}$

2. $\begin{array}{r} \frac{3}{4} = \frac{3}{4} \\ -\frac{1}{2} = \frac{\square}{4} \\ \hline \end{array}$

3. $\begin{array}{r} \frac{2}{3} = \frac{\square}{6} \\ -\frac{1}{6} = \frac{1}{6} \\ \hline \end{array}$

4. $\begin{array}{r} \frac{9}{10} = \frac{9}{10} \\ -\frac{3}{5} = \frac{\square}{10} \\ \hline \end{array}$

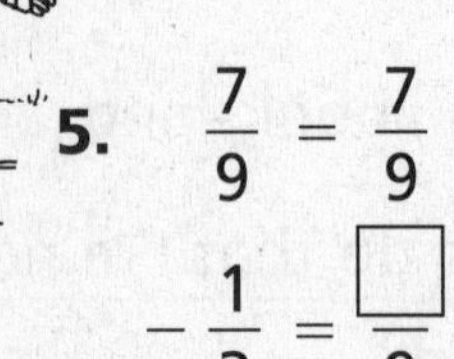

5. $\begin{array}{r} \frac{7}{9} = \frac{7}{9} \\ -\frac{1}{3} = \frac{\square}{9} \\ \hline \end{array}$

6. $\begin{array}{r} \frac{7}{8} = \frac{7}{8} \\ -\frac{3}{4} = \frac{\square}{8} \\ \hline \end{array}$

7. $\begin{array}{r} \frac{3}{5} = \frac{\square}{10} \\ -\frac{4}{10} = \frac{4}{10} \\ \hline \end{array}$

8. $\begin{array}{r} \frac{3}{4} = \frac{\square}{12} \\ -\frac{5}{12} = \frac{5}{12} \\ \hline \end{array}$

Number Sense

Directions Write a fraction below each circle to show the subtraction.

9.

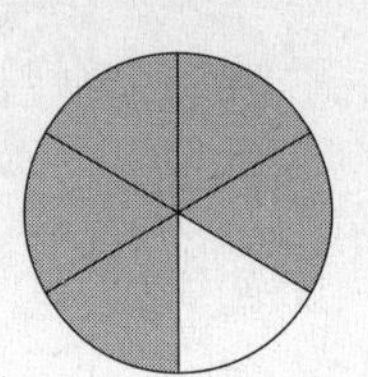

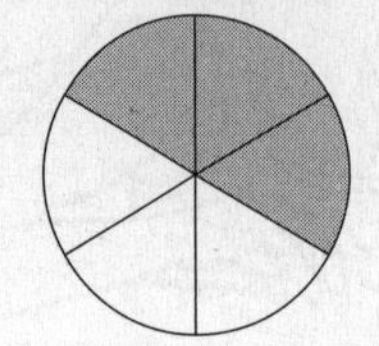

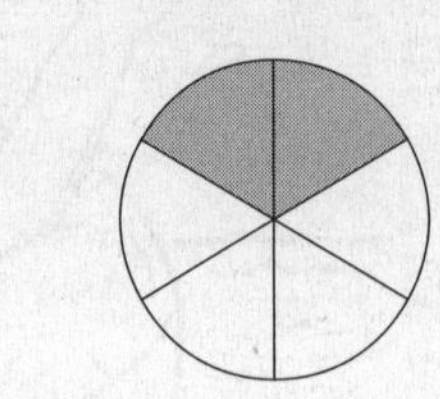

__________ − __________ = __________

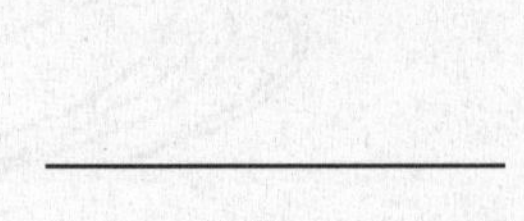

Basketball Timer

Directions **Read the word problems and solve.**

Geraldo writes down the amount of time that he spends playing basketball. His results for the first three days of the week are shown. Use the chart to solve the problems.

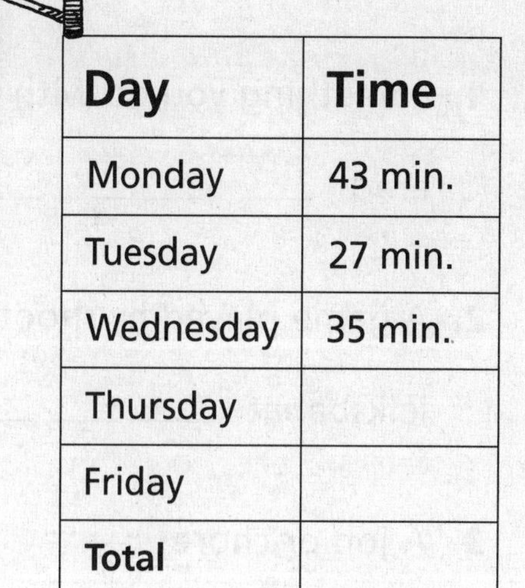

Day	Time
Monday	43 min.
Tuesday	27 min.
Wednesday	35 min.
Thursday	
Friday	
Total	

1. Estimate the amount of time Geraldo will play basketball on Thursday and Friday. Write the time on the chart for each day.

2. Estimate the total time Geraldo will spend playing basketball during the week. Write your answer in hours and minutes.

   ____

3. What assumptions did you make when you estimated total time?

   ____

   ____

4. Estimate the amount of time Geraldo spends playing basketball each month. Explain your answer.

   ____

   ____

   ____

5. Estimate the amount of time Geraldo will spend playing basketball in a year.

   ____

   ____

What To Do?

Directions **Read each definition. Then unscramble the letters to find the answers. Write the answers on the blanks. Use the numbered letters to complete the riddle.**

1. Something you do with a book

erad ___ ___ ___ ___
(blank 3 = 1)

2. A game played by shooting a ball into a hoop

lelksbabat ___ ___ ___ ___ ___ ___ ___ ___ ___ ___
(blank 9 = 2)

3. A job or chore

kowr ___ ___ ___ ___
(blank 2 = 3)

4. Something you do at a playground

wnisg ___ ___ ___ ___ ___
(blank 5 = 4)

5. A way to move when wearing shoes with wheels

etaks ___ ___ ___ ___ ___
(blank 5 = 5)

6. Something to do with a crayon

wrad ___ ___ ___ ___
(blank 2 = 6)

Now solve the riddle: Paul Bunyan was

___   ___ ___ ___ ___ ___ ___
1   2 3 4 4 5 6

Subtract

Directions Complete each subtraction problem.

1. $\frac{3}{4} = \frac{\square}{16}$
   $- \frac{5}{16} = \frac{5}{16}$

2. $\frac{1}{2} = \frac{\square}{10}$
   $- \frac{1}{10} = \frac{1}{10}$

3. $\frac{4}{5} = \frac{\square}{15}$
   $- \frac{2}{15} = \frac{2}{15}$

4. $\frac{5}{6} - \frac{5}{12}$

5. $\frac{4}{4} - \frac{5}{16}$

6. $\frac{9}{10} - \frac{2}{5}$

Use the Table

Directions Mario has downloaded five songs from the Internet. The length of each song is shown in the table. Use the table for questions 7 and 8.

Song Number	1	2	3	4	5
Number of seconds	184	224	210	304	247

7. Estimate how long it will take Mario to listen to all five songs. ____________

8. Mario has decided to purchase another three songs. Estimate the total length of all eight songs. ____________________

READING CHECK-UP

Monarch Butterflies

The monarch butterfly is an unusual insect. First, it is one of nature's most beautiful creatures. It is dressed in patterns of black and orange. Second, the butterfly can travel long distances.

When the weather becomes cold, the butterflies fly to a warm place. Some travel more than 2,000 miles!

Monarch butterflies from Canada and the cold regions of the United States often travel to California, Mexico, and Florida when the weather turns cold. They stay in the warm places all winter. They build up their energy for the long flight home in the spring.

In the spring, the butterflies travel north again. The females stop along the way to lay their eggs on leaves. Many of these females never finish the trip north. They will die along the way. The eggs that are laid will become caterpillars. The caterpillars will change into butterflies. These new butterflies continue the trip their parents began. When the weather becomes cold, these young butterflies, like their parents, travel south to the warm temperatures. Amazingly, they return to the actual trees where they were born.

Directions **Fill in the bubble next to the correct answer.**

1. Which statement is a fact?

- (A) Monarch butterflies stay in warm places all winter.
- (B) Monarch butterflies are beautiful.
- (C) Monarch butterflies are amazing creatures.
- (D) Monarch butterflies are unusual.

2. How are the young butterflies like their parents?

- (A) They travel north in the winter.
- (B) They can travel long distances.
- (C) They are born in Mexico.
- (D) They die in the winter.

3. What happens when the weather turns cold?

- (A) The butterflies die.
- (B) The butterflies travel north.
- (C) The butterflies travel south.
- (D) The females lay their eggs.

4. The author wrote this passage mainly to

- (A) explain how caterpillars turn into butterflies.
- (B) tell about the life of a monarch butterfly.
- (C) try to get you to study butterflies.
- (D) describe what the monarch butterfly looks like.

READING CHECK-UP

Blue Whales

Whales, the largest of all animals, live in oceans and seas. Whales belong to the group of animals called mammals. Mammals have a backbone and feed their babies with milk from the mother. Other mammals are dogs, horses, and humans.

There are many types of whales. The largest whale, the blue whale, is larger than an elephant and larger than the biggest dinosaur known to have lived on our planet. Blue whales average 85 feet in length and 125 tons in weight. The largest blue whale female ever seen grew to 110 feet and weighed 190 tons. That equals the weight of about 2,500 adults.

Blue whale babies, or calves, begin their lives bigger than human adults. During the last two months before they are born, blue whale calves gain two tons in weight. At birth the calf is 20 to 23 feet long. When whale calves are only 2 months old, they grow as much as nine pounds an hour.

To keep up with all this growth, the 2-month-old calf drinks about 44 gallons of milk a day. By 7 months, the calf has doubled in size. Blue whales continue to grow for the next 25 to 30 years.

Directions **Fill in the bubble next to the correct answer.**

5. The word **mammal** in the story means

Ⓐ animals that like milk.

Ⓑ animals that have a backbone.

Ⓒ animals that swim.

Ⓓ animals that have fins.

6. **WHALE** is to **CALF** as **CAT** is to

Ⓐ lion.

Ⓑ puppy.

Ⓒ kitten.

Ⓓ cow.

7. At birth, a calf is

Ⓐ 20 to 23 inches long.

Ⓑ 20 to 23 feet long.

Ⓒ as long as 2,500 people.

Ⓓ 49 feet long.

8. Which statement below is **not** a fact about blue whales?

Ⓐ Whales live in the ocean.

Ⓑ The blue whale is the largest of all whales.

Ⓒ Whales grow for 30 years.

Ⓓ Whales can drink 44 gallons of milk in one hour.

Hodja's Journey

Centuries ago in Turkey, a man named Hodja and his young son went on a long journey. Hodja gathered their few belongings and led his donkey from the stable. Hodja walked so that his son could ride.

Along the way, they came across some travelers. "Look at that young boy on the donkey," they called out. "What a lazy boy, letting his poor old father walk!" Their words make the boy feel ashamed, and he asked his father to ride. So Hodja climbed upon the donkey, and his son walked beside him.

After a while, they met more people. A man exclaimed, "The poor little boy has to walk while his father rides. What a mean man!" Now Hodja felt guilty, so he asked his son to climb upon the donkey and sit behind him.

Soon, another group of travelers saw them. "Oh, what a cruel man. How terrible that the donkey must carry two people," they groaned. Hodja decided the both should walk. "No one can complain now!" Hodja said to his son.

Directions Fill in the bubble next to the correct answer.

9. This story is mostly about

Ⓐ a poor, tired donkey on a long journey.

Ⓑ a man who tries to please everyone, yet pleases no one.

Ⓒ a man who traveled with his son.

Ⓓ a boy who feels embarrassed.

10. What is the author's main purpose in writing this story?

Ⓐ to give information about how people traveled long ago

Ⓑ to explain a lesson and persuade you to make your own decisions

Ⓒ to describe how Hodja prepared for his journey

Ⓓ to tell an amusing story

11. The story takes place in

Ⓐ Armenia.

Ⓑ Iran.

Ⓒ Africa.

Ⓓ Turkey.

READING CHECK-UP

Manny the Moose On the Loose

Alma Petty was caught by surprise Thursday night. As she washed dishes, Mrs. Petty saw a large moose run past her kitchen window. She called the local wildlife park.

Officials from the park reported that they had been taking Manny the moose to the vet for his shots when he bolted from the trailer. Officials followed leads as people called to say they had seen a moose running down the street.

The police and animal trainers were always a step or two behind Manny until Mrs. Petty called. It was Mrs. Petty's sighting that led to the capture of Manny the moose.

Mrs. Petty lives on a large lot with a swimming pool. The moose stopped at the pool for a drink. The animal trainers were able to approach Manny with a bag of treats. He followed them back into the trailer he had escaped from that morning.

Mrs. Petty, the trainers, and the police involved in the search were pleased that Manny had not been hurt. The owner of the wildlife park said this was the first time any of the animals had gotten away. He said that the vet visits most of the animals at the park. But since Manny was a new addition to the park, he had missed the vet's visit.

Directions **Fill in the bubble next to the correct answer.**

12. Why was Manny the Moose going to the vet?

- Ⓐ He was sick from eating too many treats.
- Ⓑ He needed to get his shots.
- Ⓒ He was going to live in another wildlife park.
- Ⓓ He needed a special kind of water.

13. What was special about Mrs. Petty's yard that made Manny stop there?

- Ⓐ She had a pool with water to drink.
- Ⓑ She had a yard with grass to eat.
- Ⓒ She had trees to hide in.
- Ⓓ Another moose lived there.

14. Which event came last in the selection?

- Ⓐ Manny got loose.
- Ⓑ Manny stopped to drink water.
- Ⓒ People called to report a moose running down the street.
- Ⓓ Mrs. Petty saw Manny.

15. What will probably happen next to Manny?

- Ⓐ He will go swimming.
- Ⓑ He will stay with Mrs. Petty.
- Ⓒ He will live in another wildlife park.
- Ⓓ He will go to the vet.

STOP! **Number Correct: ____ out of 15**

MATH CHECK-UP

Directions Read each question. Fill in the bubble next to the correct answer.

1. Which of these numbers is more than 1,657 and less than 1,756?

- (A) 1,576
- (B) 1,765
- (C) 1,567
- (D) 1,675

2. What is the value of 3 in 2,139?

- (A) 43
- (B) 30
- (C) 300
- (D) 3,000

3. Which number is the same as the number in the place-value chart?

1,000s	100s	10	1s
1	0	9	2

- (A) 1,002
- (B) 1,000110011011
- (C) one thousand ninety-two
- (D) 1 thousand, 9 hundreds, 0 tens, 2 ones

4. Which of these means the same as thirty-four?

- (A) 43
- (B) 340
- (C) 403
- (D) 34

5. $163 - 27 =$

- (A) 146
- (B) 36
- (C) 136
- (D) 190

6. $23 + 15 + 82 =$

- (A) 118
- (B) 130
- (C) 110
- (D) 120

7. $\begin{array}{r} 451 \\ -\ 18 \\ \hline \end{array}$

- (A) 469
- (B) 433
- (C) 443
- (D) 453

8. $18 \times 7 =$

- (A) 126
- (B) 76
- (C) 112
- (D) 25

9. $77 \div 11 =$

- (A) 171
- (B) 177
- (C) 7
- (D) 77

MATH CHECK-UP

10. Which fraction names the shaded part of the whole?

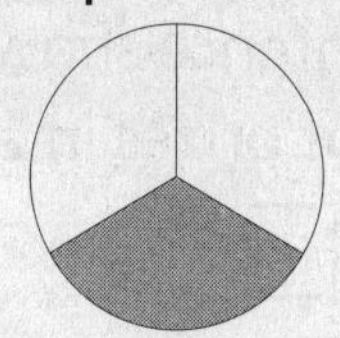

Ⓐ $\frac{1}{2}$

Ⓑ $\frac{1}{3}$

Ⓒ $\frac{1}{4}$

Ⓓ $\frac{1}{5}$

11. Which fraction words name the shaded part of the group?

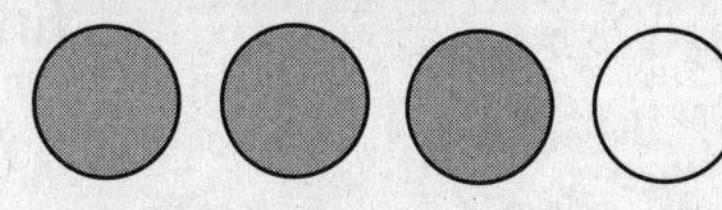

Ⓐ one quarter

Ⓑ one third

Ⓒ three fourths

Ⓓ four fourths

12. Which fraction is equivalent to $\frac{1}{2}$?

Ⓐ $\frac{1}{4}$

Ⓑ $\frac{2}{1}$

Ⓒ $\frac{2}{2}$

Ⓓ $\frac{2}{4}$

13. Write $\frac{5}{2}$ as a mixed number.

Ⓐ $\frac{2}{5}$

Ⓑ $1\frac{1}{2}$

Ⓒ $2\frac{1}{2}$

Ⓓ 3

14. $\frac{2}{5} + \frac{1}{5} =$

Ⓐ $\frac{3}{5}$

Ⓑ $\frac{3}{10}$

Ⓒ $\frac{1}{5}$

Ⓓ $\frac{1}{2}$

15. $\frac{4}{7} - \frac{2}{7} =$

Ⓐ $\frac{6}{7}$

Ⓑ $\frac{3}{7}$

Ⓒ $\frac{2}{7}$

Ⓓ 2

16. $1\frac{1}{4} - \frac{3}{4} =$

Ⓐ $\frac{1}{2}$

Ⓑ $\frac{1}{4}$

Ⓒ $\frac{3}{4}$

Ⓓ $1\frac{1}{2}$

17. $\frac{1}{2} + \frac{3}{8} =$

Ⓐ $\frac{1}{4}$

Ⓑ $\frac{3}{8}$

Ⓒ $\frac{7}{8}$

Ⓓ $\frac{4}{10}$

18. $\frac{3}{5} - \frac{1}{2} =$

Ⓐ $\frac{1}{10}$

Ⓑ $\frac{2}{3}$

Ⓒ $\frac{2}{5}$

Ⓓ $\frac{2}{7}$

19. What number goes in both boxes to make the number sentence true?

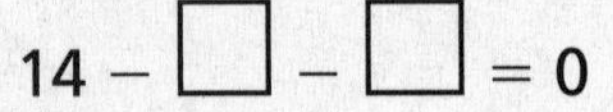

14 − □ − □ = 0

Ⓐ 0

Ⓑ 7

Ⓒ 2

Ⓓ 14

20. Ana is making surprise bags for a party. She puts 6 prizes in each of 12 bags. Which number sentence shows how to find the number of prizes in all?

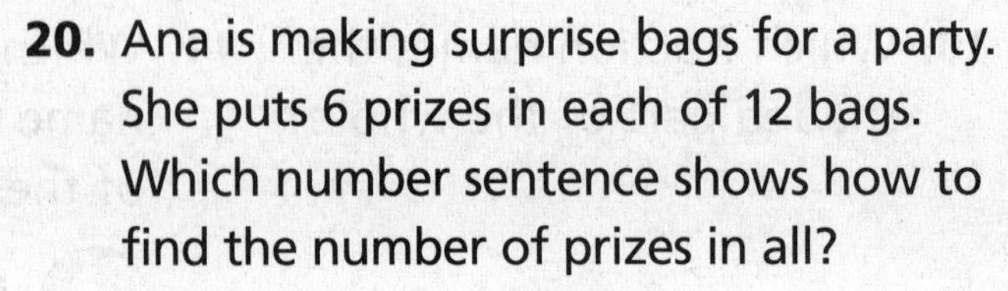

Ⓐ $6 \times 12 = \square$

Ⓑ $6 + 12 = \square$

Ⓒ $6 - 12 = \square$

Ⓓ $6 \div 12 = \square$

The pie chart shows the favorite pets of students. Study the chart. Then answer questions 21 and 22.

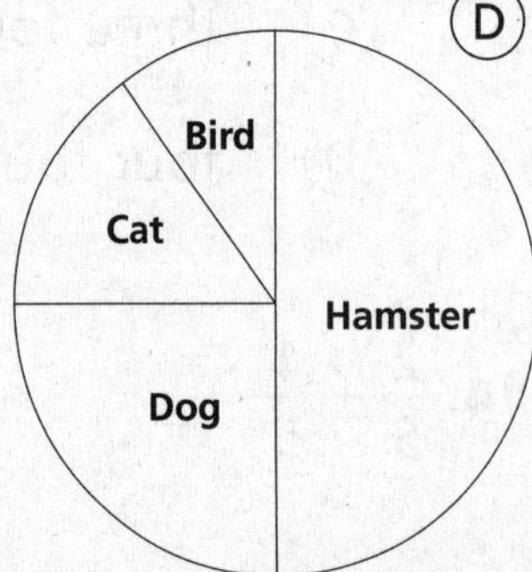

21. Which pet was chosen least?

Ⓐ cat

Ⓑ dog

Ⓒ bird

Ⓓ hamster

22. Which animal is the favorite pet?

Ⓐ cat

Ⓑ hamster

Ⓒ bird

Ⓓ dog

23. Five students are waiting for the bus. Ted is first. Rico is last. Marta is before Rico, and Amy is before Marta and after Lin. In what order are the students standing from first to last?

Ⓐ Ted, Lin, Amy, Marta, Rico

Ⓑ Ted, Marta, Lin, Amy, Rico

Ⓒ Ted, Amy, Lin, Marta, Rico

Ⓓ Rico, Marta, Amy, Lin, Ted

24. What time will the clock show in one and one quarter hours?

Ⓐ 1:45

Ⓑ 2:40

Ⓒ 2:35

Ⓓ 1:25

STOP! Number Correct: ____ out of 24

Congratulations!
Samantha
(name)
has completed Summer Counts!
Good job!
Have a good school year.

ANSWER KEY

Page 5

1. Possible answer: The parades today have big floats, marching bands, and clowns.
2. Drawings will vary but could show a girl eating pie without using her hands.
3. Answers will vary.
4. Possible answer: Red, white, and blue are the colors of the American flag.

Page 6

1. picnic-S; park-S; miles-P; school-S
2. picnic-S; bus-S
3. students-P; lunches-P;
4. teacher-S; apples-P; jug-S; lemonade-S
5. park-S; pond-S; geese-P; ducks-P
6. class-S; walk-S; trail-S

Page 7

1. chicken
2. teacher
3. watch
4. which
5. rich
6. charge
7. such
8. teacher
9. lunch
10. chicken
11. peach
12. children
13. watch
14. reach
15. inches

Page 8

Estimates may vary.

2. 60 + 30 = 90; 88
3. 80 + 40 = 120; 125
4. 200 + 40 = 240; 225
5. 200 + 100 = 300; 334
6. 600 + 700 = 1,300; 1,365
7. $5.00 + $2.00 = $7.00; $6.88
8. $6.00 + $3.00 = $9.00; $9.18
9. $27.00+$11.00= $38.00; $38.45

Page 9

1. 0, 2, 4, 6, 8
2. **a.** Numeral order may vary. 84+62=146 or 82+64=146
   b. 146
3. **a.** Numeral order may vary. Possible answers: 30 − 29; 40 − 39
   b. 1

Page 10

Park: 4th row down, 8th letter across. Down.
Lunch: 4th row down, 2nd letter across. Diagonally.
Chicken: 9th row down, 4th letter across. Across.
Coolers: 1st row down, 3rd letter across. Diagonally.
Family: 1st row down, 4th letter across. Across.
Lemonade: 3rd row down, 2nd letter across. Across.
Class: 4th row down, 1st letter across. Diagonally.
Laugh: 4th row down, 6th letter across. Down.
Ants: 10th row down, 3rd letter across. Across.
Picnic: 2nd row down, 1st letter across. Down.
Basket: 5th row down, 9th letter across. Down.
Playground: 1st row down, 10th letter across. Down.

Page 11

1. D
2. C
3. B
4. B

Page 13

1. A spider hunts by running after a victim and attacking it. It hides and waits for a victim to pass by. It makes a trap to catch a victim.
2. A spider uses venom to kill its victims.
3. A spider has fangs instead of teeth.
4. Possible answers: lizards and snakes
5. Real. It tells facts about spiders.

Page 14

1. elephant's
2. teacher's
3. animals'
4. monkey's
5. Sarah's
6. ducks'
7. Gilberto's
8. butterfly's

Page 15

1. animal
2. mouse
3. kitten
4. wool; turtle
5. wild; cold
6. dragon
7. cattle
8. bear
9. kitten
10. mouse

Page 16

Estimates may vary.

2. 600−300=300; 232
3. 900−500=400; 364
4. 300−100=200; 212
5. 700−200=500; 472
6. 400−300=100; 89
7. $6.00−$1.00=$5.00; $4.72
8. $8.00−$3.00=$5.00; $5.05
9. $10.00−$8.00=$2.00; $1.51

Page 17

1. spring
2. fall
3. 68
4. 48
5. 16
6. **a.** 95
   b. 69
7. Day 10
8. 14

Page 18

1. spider
2. animals
3. fangs
4. wool
5. hole
6. prey

Riddle: a spider glider

Page 19

1. C
2. A
3. C
4. D

Page 21

1. in Owen Foote's bedroom
2. Owen likes to read. He has books all over his bed.
3. A silverback is the leader of the gorillas.
4. They liked the silverback and wanted to be around him.
5. Owen is proud. He feels like somebody.

Page 22

1. Common: team; Proper: July
2. Common: bus; Proper: Bailey Street
3. Common: game; Proper: Texas
4. Common: day, trip, bus; Proper: Wyoming
5. Common: mountains; Proper: Colorado
6. Common: friend; Proper: Manuel
7. Common: game; Proper: Joe Hollings Field
8. Common: team, game; Proper: Thursday

Page 23

1. visit
2. throw
3. hurt
4. remember
5. basketball
6. bats
7. softball

8. home runs
9. swim
10. underwater
11. kick
12. Sports

Page 24

Estimates may vary.

2. 30×4=120; 128
3. 60×5=300; 305
4. $2.00×2=$4.00; $4.48
5. $3.00×4=$12.00; $12.08
6. $8.00×5=$40.00; $41.70
7. 100×7=700; 798
8. 300×5=1,500; 1,405
9. 500×8=4,000; 4,216

Page 25

Methods to solve the problems may vary.

1. 8 laps. 20 minutes is twice as long as 10 minutes. So Lucia will swim 4 × 2, or 8 laps in 20 minutes.
2. 24 laps. 1 hour equals 60 minutes. 60 minutes is six times as long as 10 minutes. Lucia swims 4 × 6, or 24 laps in 1 hour.
3. 40 minutes. It took Lucia 10 minutes to swim 4 laps. 16 laps is 4 times greater than 4 laps. It will take 4 times as long to swim 4 laps, or 4 × 10 = 40.
4. 50 days. Lucia wants to swim 200 laps at 4 laps a day. 200÷4=50, or the number of days it will take to swim 200 laps.

Page 26

Across:

1. baseball
3. skateboard
4. pitcher
6. team

Down:

1. basketball
2. soccer
4. points
5. bat

Page 27

Estimates may vary.

1. 5,000 × 8 = 40,000
2. 500 × 6 = 3,000
3. 900 × 7 = 6,300
4. 1,000 × 4 = 4,000
5. 6,000 × 3 = 18,000
6. 4,000 × 2 = 8,000
7. 8,000 × 5 = 40,000
8. 3,000 × 9 = 27,000
9. 5,000 × 3 = 15,000
10. 8,000 × 8 = 64,000
11. 5,000 × 4 = 20,000
12. 3,000 × 7 = 21,000
13. 3,000 × 6 = 18,000
14. 7,000 × 2 = 14,000

Page 29

1. Possible answers: paint, glitter, ribbon
2. Answers will vary, but should include steps listed on page 28.
3. Answers will vary.
4. If you don't follow the directions in order, the card will not turn out right.

Page 30

1. one line: was; two lines: making
2. one line: are; two lines: going
3. one line: is; two lines: turning
4. one line: had; two lines: made
5. one line: will; two lines: draw
6. one line: is; two lines: helping
7. one line: has; two lines: given
8. one line: will; two lines: make

Page 31

1. new
2. board
3. cent
4. heard
5. write
6. wood
7. write
8. would
9. guess
10. because
11. Please
12. letter
13. again
14. instead

Page 32

2. 7 r4
3. 9 r4
4. 158
5. 122
6. 159 r1
7. $2.05
8. $2.96
9. $2.37

Page 33

1. 2; 3; 5
2. A striped section because that pattern covers a greater part of the spinner.
3. more than 50; The striped areas are the larger areas.

Page 34

Card: 1st row down, 4th letter across. Diagonally.
Letter: 1st row down, 5th letter across. Across.
Message: 2nd row down, 10th letter across. Down.
Mystery: 1st row down, 2nd letter across. Diagonally.
Scissors: 2nd row down, 1st letter across. Diagonally.
Envelope: 1st row down, 9th letter across. Down.
Read: 2nd row down, 8th letter across. Down.
Mail: 2nd row down, 4th letter across. Across.
Write: 8th row down, 1st letter across. Across.
Stamps: 2nd row down, 1st letter across. Down.
Address: 4th row down, 1st letter across. Diagonally.
Markers: 10th row down, 1st letter across. Across.

Page 35

1. 325
2. 1,245
3. 1,165
4. 307
5. 982
6. 3,144
7. blue
8. Answers will vary.
9. Answers will vary.

Page 37

1. B
2. A
3. Answers will vary.

Page 38

1. went
2. borrowed
3. bought
4. decided
5. wanted
6. knew
7. spent
8. filled
9. packed
10. loaded; headed

Page 39

Sentences for 1–3 will vary. Accept reasonable responses.

1. desert
2. cave
3. cactus
4. Visitors
5. streets
6. beach
7. cactus
8. forest
9. mountain
10. river
11. cave

Page 40

1. **a.** part; whole
   b. single object; group
2. **a.** equal parts being described
   b. equal parts the whole is divided into

Page 41

1. **a.** $\frac{1}{2}$; **b.** $\frac{1}{3}$; **c.** $\frac{1}{4}$; **d.** $\frac{1}{5}$
2. $\frac{3}{8}$; three-eighths
3. $\frac{7}{10}$; seven-tenths

4. $\frac{3}{5}$; three-fifths
5. $\frac{4}{7}$; four-sevenths

Page 43

1. B
2. D
3. D
4. C

Page 45

1. It is an instrument used by doctors to listen to the heart and lungs.
2. It helped doctors know what was happening inside the body.
3. Possible answers: ears, eyes, heart, and lungs.
4. Because they could not see inside the body to find out why someone was sick.
5. Real. It tells facts about a stethoscope, organs, and doctors.

Page 46

1. will conduct
2. will choose
3. will give
4. will have
5. will exhibit
6. will decide
7. will award
8. will visit
9. will read
10. will win

Page 47

1. knuckle
2. wrist
3. ears
4. head
5. skeleton
6. bones
7. skull
8. hair
9. face
10. eyes
11. mouth
12. teeth

Page 48

1. Answers will vary.
2. Circle: $\frac{1}{8}$, $\frac{11}{16}$, $\frac{3}{4}$
3. $1\frac{1}{2}$; $1\frac{1}{4}$; $1\frac{4}{8}$; $1\frac{3}{16}$
4. $\frac{5}{3}$, $\frac{9}{5}$, $\frac{19}{8}$, $\frac{27}{7}$

Page 49

1. a. Including Homer, everyone will get 6 pretzels.
2. Each person will get $4\frac{1}{2}$ pretzels.
3. Pizza should have 16 equal lines drawn on it.
4. 2 pieces
5. 4 pieces

Page 50

Across:

3. proper
4. stethoscope
6. teeth
8. skull

Down:

1. skeleton
2. future
5. organs
7. health

Page 51

1. B
2. A
3. C
4. D

Page 53

1. Answers will vary.
2. Possible answers: Sydney's offers all you can eat pizza on Tuesday and free salad on Friday.
3. Answers will vary.

Page 54

1. He
2. them
3. it
4. him
5. She
6. they
7. We
8. they
9. us
10. it

Page 55

1. f
2. a
3. e
4. c
5. d
6. b
7. table
8. dinner
9. dishes
10. water
11. drink
12. sugar
13. salt
14. bread
15. picnic
16. grill

Page 56

1. $\frac{3}{4}$
2. $\frac{4}{5}$
3. $\frac{7}{6}$
4. $\frac{8}{8}$; 1
5. $\frac{2}{3}$
6. $\frac{4}{5}$
7. $\frac{5}{6}$
8. $\frac{9}{8}$
9. $\frac{13}{10}$

Page 57

1. Christy's and the Italian Kitchen
2. \$9.75 to \$14.00, or \$14.00 – \$9.75 = \$4.25.
3. a. King Pizza, Mama Mia's, Heavenly Pizza, Pizza House, Christy's, Italian Kitchen, Luigi's
   b. Pizza House

Page 58

My favorite food is pizza. The pizza has to have lots of cheese and pepperoni.

Page 59

1. D
2. C
3. B
4. C

Page 61

1–4. Answers will vary. Accept reasonable responses.

Page 62

1. excited
2. famous
3. old
4. helpful
5. knowledgeable
6. thrilled
7. graceful
8. exciting
9. long
10. happy, tired

Page 63

1. color
2. stories
3. music
4. museum
5. song
6. sing
7. stories
8. paintings
9. oil
10. pencil
11. art
12. music

Page 64

1. $\frac{1}{2}$
2. $\frac{2}{5}$
3. $\frac{2}{3}$
4. $\frac{7}{8}$
5. $\frac{1}{3}$
6. $\frac{3}{5}$
7. $\frac{1}{3}$
8. $\frac{1}{4}$
9. $\frac{3}{4} - \frac{1}{4} = \frac{2}{4}$ or $\frac{1}{2}$
10. $\frac{1}{4} - \frac{1}{2} = \frac{3}{4}$

Page 65

1. **Library:** 2 boxes down, 6 boxes across.
2. **Museum:** 4 boxes down, 9 boxes across.
3. **Theater:** 2 boxes down, 4 boxes across.
4. **Art store:** 6 boxes down, 6 boxes across.

5. **Craft store:** 7 boxes down, 9 boxes across.
6. Go 6 blocks north and 4 blocks east.
7. Go 7 blocks west and 5 blocks south.

Page 67

1. $\frac{3}{5}$
2. $\frac{2}{5}$
3. $\frac{1}{4}$
4. $\frac{1}{2}$
5. $\frac{4}{9}$
6. $\frac{1}{4}$
7. $\frac{2}{7}$
8. $\frac{1}{7}$
9.

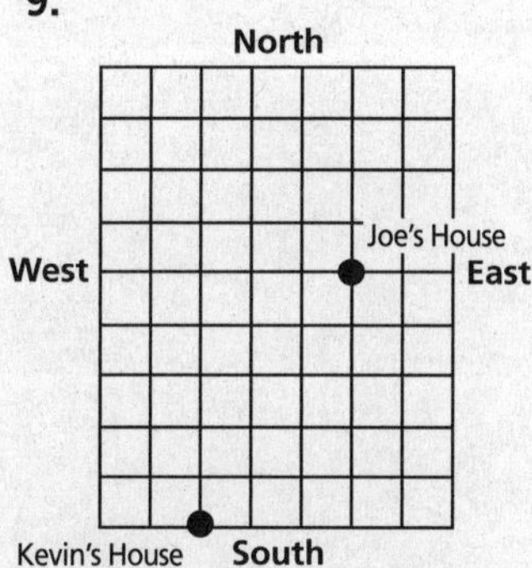

10. Go north for five blocks. Then go east for three blocks.
11. Go west for three blocks. Then go south for five blocks.

Page 69

1. The fifth of May
2. The Mexican army's victory over the French army.
3. That's when the fun starts!
4. Answers will vary. Accept reasonable responses.
5. Real. It tells about a battle, a holiday, and how the holiday is celebrated.

Page 70

1. question
2. statement
3. command
4. question
5. exclamation

Page 71

1. afternoon
2. everyone
3. maybe
4. without
5. grandmother
6. birthday
7. sometime
8. breakfast
9. Everyone
10. airplane

Page 72

2. 2
3. 4
4. 10
5. 4; $\frac{5}{8}$
6. 2; $\frac{5}{8}$
7. 4; $\frac{5}{6}$
8. 6; $\frac{7}{10}$
9. $\frac{2}{8} = \frac{3}{8} = \frac{5}{8}$

Page 73

1. **a.** $\frac{3}{8}$; **b.** $\frac{5}{8}$
2. **a.** 80; **b.** 9
3. **a.** $\frac{2}{4}$ hour
   b. 14 hour
   c. 2:30 p.m.

Page 74

1. battle
2. army
3. colorful
4. perform
5. great
6. guitars

Page 75

1. $\frac{5}{10}$
2. $\frac{6}{9}$
3. $\frac{4}{8}$
4. $\frac{8}{10}$
5. $\frac{5}{10} + \frac{3}{10} = \frac{8}{10} = \frac{4}{5}$
6. $\frac{6}{9} + \frac{4}{9} = \frac{10}{9} = 1\frac{1}{9}$
7. $\frac{6}{10} + \frac{3}{10} = \frac{9}{10}$
8. $\frac{14}{15}$
9. $\frac{7}{10}$
10. $\frac{1}{2}$
11. $\frac{11}{12}$
12. $\frac{5}{8}$
13. $\frac{7}{9}$

Page 77

1. Possible answer: He straightened a road for the settlers.
2. To make room for farms and towns
3. He was sad. The settlers would not have shade. They would not hear the leaves rustle.
4. They will plant small trees in the Dakotas.
5. He said this to Babe.

Page 78

1. harder
2. strongest
3. more often
4. more closely
5. more loudly
6. sooner

Page 79

1. write
2. fly
3. read
4. run
5. bore
6. walk
7. work
8. swim
9. swimming
10. study
11. running
12. flying
13. draw
14. reading
15. writing
16. bored

Page 80

1. 4; $\frac{1}{8}$
2. 2; $\frac{1}{4}$
3. 4; $\frac{3}{6}$
4. 6; $\frac{3}{10}$
5. 3; $\frac{4}{9}$
6. 6; $\frac{1}{8}$
7. 6; $\frac{2}{10}$
8. 9; $\frac{4}{12}$
9. $\frac{5}{6} - \frac{3}{6} = \frac{2}{6}$ or $\frac{1}{3}$

Page 81

1. Thursday: 35 min.; Friday: 35 min.
2. 2 hours and 55 minutes
3. I estimated he would play for 35 minutes on Thursday and Friday.
4. 700 minutes. I multiplied 175 minutes a week by four weeks.
5. 9100 minutes. There are 52 weeks in a year. I multiplied 175 minutes a week by 52 weeks.

Page 82

1. read
2. basketball
3. work
4. swing
5. skate
6. draw

Riddle: a logger

Page 83

1. $\frac{12}{16} - \frac{5}{16} = \frac{7}{16}$
2. $\frac{5}{10} - \frac{1}{10} = \frac{4}{10} = \frac{2}{5}$
3. $\frac{12}{15} - \frac{2}{15} = \frac{10}{15} = \frac{2}{3}$
4. $\frac{5}{12}$
5. $\frac{11}{16}$
6. $\frac{1}{2}$
7. Estimates will vary. Actual answer: 1,169
8. Estimates will vary.

Page 84

1. A
2. B
3. C
4. B

Page 85

5. B
6. C
7. B
8. D

Page 86

9. B
10. B
11. D

Page 87

12. B
13. A
14. B
15. D

Page 88

1. D
2. B
3. C
4. D
5. C
6. D
7. B
8. A
9. C

Page 89

10. B
11. C
12. D
13. C
14. A
15. C
16. A
17. C
18. A

Page 90

19. B
20. A
21. C
22. B
23. A
24. C